Biomaterial-supported Tissue Reconstruction or Regeneration

Edited by Mike Barbeck, Ole Jung, Ralf Smeets and Tadas Koržinskas

Published in London, United Kingdom

IntechOpen

Supporting open minds since 2005

Biomaterial-supported Tissue Reconstruction or Regeneration
http://dx.doi.org/10.5772/intechopen.77756
Edited by Mike Barbeck, Ole Jung, Ralf Smeets and Tadas Koržinskas

Contributors
Maria Fulgência Bandeira, Nikeila Conde, Carina Toda, Ana Regina Casaroto, Leilane Bentes, Geisy Lima, Samuel Lungareze, Larissa Silva, Joyce Meira, Marne Vasconcellos, Zeljka Peric Kacarevic, Andrea Ruffini, Simone Sprio, Anna Tampieri, Lorenzo Preti, Masaru Murata, Bano Subia, Ujjwal Dahiya, Sarita Mishra, Rosalba Parenti, Anna Dolcimascolo, Giovanna Calabrese, Sabrina Conoci, Elena Fuentes, Sofia Alves, Ainara López, Lucía Mendizabal, Virginia Sáenz De Viteri

First published in London, United Kingdom, 2019 by IntechOpen
IntechOpen is the global imprint of INTECHOPEN LIMITED, registered in England and Wales, registration number: 11086078, The Shard, 25th floor, 32 London Bridge Street
London, SE19SG – United Kingdom
Printed in Croatia

British Library Cataloguing-in-Publication Data
A catalogue record for this book is available from the British Library

Additional hard copies can be obtained from orders@intechopen.com

Biomaterial-supported Tissue Reconstruction or Regeneration
Edited by Mike Barbeck, Ole Jung, Ralf Smeets and Tadas Koržinskas
p. cm.
Print ISBN 978-1-83880-377-3
Online ISBN 978-1-83880-378-0
eBook (PDF) ISBN 978-1-83880-433-6

For EU product safety concerns:
IN TECH d.o.o., Prolaz Marije Krucifikse Kozulić 3, 51000 Rijeka, Croatia, info@intechopen.com or visit our website at intechopen.com.

Meet the editors

Mike Barbeck has 17 years of experience in regenerative biomedical research working at institutes in Germany. He works on research into the principles of biomaterial-mediated tissue regeneration, which has resulted in 62 papers in peer-reviewed journals and a variety of congress distributions. His research has led to further elucidation of cellular fundamentals of the foreign body response to different biomaterials with special focus on the role and differentiation of macrophages and multinucleated giant cells for tissue regeneration. His key interests are tissue regeneration, bone regeneration, bone substitute materials, collagen-based biomaterials, inflammatory cell/tissue responses to biomaterials, degradation of biomaterials, tissue responses to biomaterials, macrophages and wound healing, and biocalcification.

Ole Jung studied medicine at the University Medical Center Hamburg-Eppendorf with stays abroad in the USA, Great Britain, Poland, and Switzerland. He finished his medical doctorate on biodegradable magnesium implants in 2016. Since 2011, he has been conducting research on magnesium materials for medical applications, biocompatible coatings, and dental implants.

Ralf Smeets, MD, DDS, PhD, is the Vice Head of the Department of Oral and Maxillofacial Surgery and the Head of the Division of "Regenerative Orofacial Medicine" at the University Medical Center Hamburg-Eppendorf, Germany. His academic career started in Aachen at RWTH Aachen University, where he completed his chemical studies in 1995 and studied medicine until 2003. Besides his medical studies, he studied economics at the University of Hagen. He finished his residency as an oral- and maxillofacial surgeon and oral surgeon in 2009. In 2011, he became a full W2 professor at the Department of Oral and Maxillofacial Surgery, University Medical Center Hamburg-Eppendorf, and was a visiting professor at University Bremerhaven (Germany) until 2015. Recently, he has been awarded an exclusive W3 Heisenberg Professorship at the University Medical Center Hamburg-Eppendorf and is the Scientific Director of the curriculum "Dental Regeneration Specialist."

Tadas Koržinskas is an oral surgeon and a certified specialist in esthetic, reconstructive dentistry and temporomandibular disorder (TMD), with 15 years of experience in oral implantology. As an international lecturer, he is recognized in the field of implant-supported restorations and TMD. His research focuses on dental implants, biomaterials, and dental occlusion.

Contents

Preface

A broad variety of biomaterials is available for different clinical approaches. Moreover, different developments in the field of biomaterials have recently been conceived. This book provides a comprehensive overview of available biomaterials and future trends in biomaterial development. Furthermore, cell and tissue reactions and related regenerative pathways of different classes of biomaterials in regenerative medicine are described. Finally, the use of biomaterials for tissue engineering strategies is described, including the different cells and the application of stem cells. Altogether, the present book should give insights into the different biomaterial classes used for tissue regeneration and the beneficial effects of newly developed materials.

The authors of this book are experts of their respective contents. Thus, the book will provide interested readers with comprehensive knowledge of many different biomaterials. Personally, as the editor, I have learned a great deal on each subject and expect the readers will also learn many new facts of the evolving field of biomaterials. Please enjoy its broad variety!

Mike Barbeck, PhD
University Medical Center Hamburg-Eppendorf,
Department of Oral Maxillofacial Surgery,
Division for Regenerative Orofacial Medicine,
Hamburg

Section 1

Biomaterial Processing Techniques

Chapter 1

Synthesis of Nanostructured Hydroxyapatite via Controlled Hydrothermal Route

Andrea Ruffini, Simone Sprio, Lorenzo Preti and Anna Tampieri

Abstract

Hydroxyapatite represents the natural inorganic component of the bone and may be considered an essential element required for the development of bone substitutes in the field of regenerative medicine. Hydroxyapatite bone substitutes own biomimetic, osteoconductive, and osteoinductive properties thanks to their chemical-physical properties and nanostructure that play a critical role for the reconstruction of calcified tissues. Controlling the structure of hydroxyapatite nanocrystals is vital for obtaining a sustained product, and it should be an advantage on the final materials suitable for bone replacement, in terms of adsorptive activity, drug delivery system, etc. Compared to other synthesis techniques, hydrothermal processing (refers to a synthesis in aqueous solution at elevated pressure and temperature, in a closed system) has the ability to precipitate the hydroxyapatite from overheated solution, regulating the rate and uniformity of nucleation, growth, and maturation, which affect size, morphology, and aggregation of the crystals. This chapter wants to provide an overview of realization of nanosized hydroxyapatite-based bioceramics (e.g., powder and 3D structures) with desired morphology of crystallites, by hydrothermal processing. In this way, some critical hydrothermal parameters fundamental on the control of the crystal shape and dimension (pH, temperature, starting precursors, etc.) are discussed.

Keywords: hydrothermal synthesis, nanostructured hydroxyapatite, crystal growth, morphology control, regenerative medicine

1. Introduction

Calcium phosphates (CaP) are the main mineral constituent of human bones and teeth. For this reason, synthetically CaP-based materials nowadays are the most ubiquitous family of biomaterials for their use in biological applications and tissue engineering. These attractive biomedical materials possess excellent biocompatibility, osteoconductive properties, nontoxicity, and chemical similarity to the inorganic component of the natural bone [1].

The realization of CaP biomaterials reproducing the calcified tissue (dense and porous block, granules, and powders) is clinically needed as an alternative to autologous- and heterologous-derived scaffolds [2]. The majority of CaP biomaterials shall be applicable for bone reconstruction and replacement in tissue engineering when the bone has no self-regenerative capacity following severe illness or trauma, as well as other applications, like drug delivery agents, prosthetic coatings, and gene carriers [3, 4].

Among the CaP family compounds, hydroxyapatite $Ca_5(PO_4)_3OH$ (HA) is the most extensively used in medicine for implant fabrication as an alternative to the human bone; it is the thermodynamically most stable phase in physiological conditions and owns the most similarity in mineralogical phase and chemical composition to the mineral part of the bone tissue [5]. The in vivo formation of this mineral occurs through the biomineralization process, where nanometer-sized crystals of HA are precipitated on collagen fibrils into mineralized self-assembled hierarchical and calcified structure. The thickness of the HA crystals ranges from about 5 to 20 nm, while the length from 15 to 200 nm [6].

Biological HA is a nonstoichiometric, carbonated, and calcium-deficient form of apatite (Ca/P atomic ratio lower than 1.67), containing various amounts of positively charged ions (e.g., Mg^{2+}, Na^+ and K^+) and negatively charged ions (e.g., CO_3^{2-}, Cl^- and F^-), in substitution of Ca^{2+} or PO_4^{3-}/OH^- ions, respectively [7, 8].

Present-day researches concern new route or improving preexisting methods to accurately engineer HA-based materials with characteristics closer to the living bone, aiming more effective applications in the field of biomaterials. The in vivo and in vitro performance of HA biomaterials remarkably depends on the development of their properties during the manufacturing process, such as microscopic characteristic (e.g., grain size topography, particles size distribution, nanostructure), morphology (e.g., porosity, pore size, 3-D architecture), chemical compositions, crystallographic structure, etc.

It is well known that biological and mechanical properties of biomaterials are strongly affected by its nanostructural characteristics. Compared to conventional ceramic formulations, the nanophase of HA materials can significantly affect their mechanical strength and the solubility that has a substantial effect on resorption and bioactivity. Furthermore, nanostructures can enhance osteoblast adhesion and affect the surface wettability for the selective control of protein interactions [9]. The mechanical properties and microstructures of the resulting HA ceramics are mostly influenced by the microstructure of the produced powder, including crystallinity, agglomeration, stoichiometry, and substitutions and the processing conditions [10].

The control over HA crystallization through a precise control of crystal nucleation and growth is a major challenge in the synthesis of crystalline particles, with defined geometry, morphologies, orientations, sizes, and composition. These intrinsic features are closely related to their properties and may affect their applications.

Many methods have been proposed in the literature to prepare nanostructured HA materials (e.g., nanoparticles or 3-D scaffolds with various shapes and sizes), such as coprecipitation, sol-gel synthesis, mechanical milling, hydrothermal reaction, etc. [11]. Between them, hydrothermal synthesis method allows more choice of variable factors that affect the morphology of the final material, such as nature of precursors and their concentration, saturation, temperature, pH, process time, and the presence of potential agents used for controlling the final morphological structures [12, 13].

This chapter focuses on the synthetic hydrothermal strategy employed in the preparation and design of different HA nanoparticles and nanostructured materials and reviews about the roles of important parameters on the HA nanostructured realization.

2. Nanostructured hydroxyapatite

CaP exhibit different characteristics as Ca/P atomic ratios, crystal systems, and solubility (in physiological conditions, T = 37°C, pH = 7.3) [14, 15]; some most common CaP used for biomedical applications are summarized in **Table 1**. CaP materials which have Ca/P atomic ratio external to 1–2 range are not suitable for implantation into the body because of their high acidity (low Ca/P), basicity (high Ca/P), and solubility. As a result, among the known CaP compounds, OCP, α/β-TCP and CDHA/HA are significantly most useful for biomedical applications, like orthopedic surgery and dentistry.

The different crystal morphologies of CaP exhibited in the various biological hard tissue have addressed the research to develop these materials with high control of nanostructural properties, as particle size, nano-surface, dimensional anisotropy, etc. For example, in the bone tissue, the HA crystallites have rod shape owning dimension of about 50 × 25 × 4 nm; in contrast, in the tooth enamel, larger hexagonal prism crystallites have the dimension of about 100 × 70 × 25 nm, with c-planes that are preferentially parallel to the collagen fibrils or enamel surface, respectively [16]. For this reason, crystals exhibit enhanced adsorption properties because of their higher charging surface and with plate-like and hexagonal morphologies, may find promising application in dental implants.

Lee et al. reported enhanced bioactivity of nanostructured HA compared to sintered and coarser ceramics: nanophase HA promotes the osteogenic differentiation of periodontal ligament cells and more efficiency of osteoclast-like cell adhesion [7]. Besides, HA nanoparticles are used for cell targeting, gene transfecting, and drug delivery thanks to their strong molecular adsorption property and increased surface area [17]. The size of HA particles also play critical roles in biological response, including cell proliferation modification, oriented cell differentiation, and cell apoptosis [18]. HA nanowires and nanosheets are capable for moderated reinforcement of the biomaterials and can be used as mechanical components to stiffen isotropic composite materials. In the case of sinterability of bioceramics, nanoparticles exhibit improved feature if compared to coarser particles [19].

Stoichiometric HA has typically hexagonal crystal system with the $P6_3/m$ space group and two principal crystal planes: a-plane and c-plane (see **Figure 1**); its crystal unit cell is characterized by a = b = 0.942 nm and c = 0.688 nm, α = β = 90° and γ = 120°. Positively charged calcium ions (Ca^{2+}) are mainly present in the a-planes,

Calcium phosphates	Acronym	Ca/P ratio	Crystal system	K_{ps} (log)
Monetite, $CaHPO_4$	DCPA	1	Triclinic	7.02
Brushite, $CaHPO_4 \cdot 2H_2O$	DCPD	1	Monoclinic	6.63
Amorphous calcium phosphate, $Ca_xH_y(PO_4)_z \cdot nH_2O$; $3 < n < 4.5$	ACP	1.2–2.2	—	<4
Octacalcium phosphate, $Ca_8(HPO_4)_2(PO_4)_4 \cdot 5H_2O$	OCP	1.33	Triclinic	95.9
α-Tricalcium phosphate, α-$Ca_3(PO_4)_2$	α-TCP	1.5	Monoclinic	25.5
β-Tricalcium phosphate, β-$Ca_3(PO_4)_2$	β-TCP	1.5	Rhombohedral	29.5
Calcium-deficient hydroxyapatite, $Ca_{5-x}(HPO_4)_x(PO_4)_{3-x}(OH)$; $0 < x < 0.5$	CDHA	1.5–1.67	Hexagonal	<42.6
Hydroxyapatite, $Ca_5(PO_4)_3(OH)$	HA	1.67	Hexagonal	58.6
Tetracalcium phosphate, $Ca_4(PO_4)_2O$	TTCP	2	Monoclinic	37–42

Table 1.
Main CaP compounds used for biomedical applications.

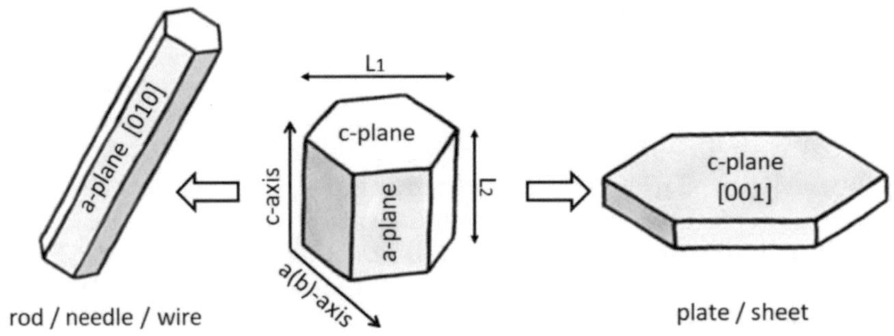

Figure 1.
Hexagonal crystal structure of stoichiometric HA, where a-plane is Ca site (positive charge) and c-plane is PO_4 site (negative charge).

and hence negatively charged phosphate (PO_4^{3-}) and hydroxide (OH^-) ions are present in the c-planes. That is, HA surfaces present anisotropic characteristics such as adsorption profiles for electrolytes or biomolecules [16].

3. Hydrothermal synthesis of calcium phosphates

The preparation of HA crystals with controlled morphology and phase pure (both in powder or scaffold forms) has been the object of intensive research over the last decades. Sadat-Shojai and Earl et al. [13, 20] summarized three main processing routes (which include specific techniques) developed for synthesizing nanosized HA:

- Dry methods (i.e., mechanochemical, solid-state reaction processes)
- Wet methods (i.e., precipitation, hydrolysis, sol-gel, hydrothermal, solvothermal, emulsion, sonochemical, microwave, chemical vapor deposition processes)
- High-temperature processes (i.e., combustion preparation, synthesis from biogenic sources)

Wet chemistry techniques are widely used for their particularly effective control of particle size, morphology, and crystallinity. Among this, the hydrothermal technique is a promising and convenient way to easily control the reaction conditions, through a one-step synthesis of a desired phase under gentle reaction conditions [21].

Byrappa and Yoshimura defined hydrothermal processing as any heterogeneous chemical reaction developed in a closed system, in the presence of an aqueous solvent and above to its boiling temperature and pressure ($T > 100°C$; $P > 1$ atm) and able to dissolve and recrystallize materials that are relatively insoluble under normal conditions. Other authors established hydrothermal conditions as a chemical precipitation in which the dissolution, crystal growth, and aging steps are conducted at a high temperature and pressure (typically above the water boiling point) inside an apparatus consisting of a steel pressure vessel called reactor/autoclave [21, 22]. Classically, a temperature gradient is maintained between the opposite ends of the vessel by using an external heater jacket with controlled temperature, and the pressure is measured by a gas pressure transmitter [23].

The hydrothermal technique can be effective in processes aimed to the production of highly homogeneous and monodispersed nanoparticles but also used as attractive techniques for processing of nanostructured 3D materials. Specifically, recent studies in hydrothermal conditions were carried out on the realization of HA ceramic generated, partially or entirely, starting from various biogenic sources, such as marine algae, shell, corals, or wood [24–27]. The biomorphic transformations in hydrothermal conditions enable to create a material with the preservation of natural-starting 3D structure (similar to that human bone) and appropriate nanostructural characteristics, use as a bone graft substitute for bone implants [28].

The achieving of a precise controlling of morphology, geometry and size, over the crystallographic and chemical structure of HA in hydrothermal conditions is related with a largest range of useful work conditions, in particular the temperature. The literature reports many variety of hydrothermal methods for the realization of HA-nanostructured materials, in a wide range of dimension and variety of morphological features of the crystals, for example, rods [20, 29–46], needles [29, 30, 32, 47–53], wires [18, 54–57], whiskers/fibers [23, 47, 49, 58–60], sheets [55, 61, 62], plates [23, 44, 47–49, 54, 63, 64], and organized rod spheres [18, 31, 44, 54, 65, 66] or sheet spheres [19, 61, 67, 68], etc. (see **Table 2** and **Figure 2**). Aldrich and Smith defined a geometric nomenclature for nanoparticles [69], frequently described in ambiguous way in many scientific works; for example, the geometry indicating elongated HA as rod, needle, and fibers should be named according to well-defined dimensional aspect ratio and c-axial length.

This study is mainly focused on the correlation between the different HA morphologies and stoichiometries obtained and the various parameters that affect crystallization in the hydrothermal process, with their complex relationship. Appropriate operating conditions can be carried out and adjustment to induce in the crystallization pathway and produce the desired morphology [70, 71]. Size and morphology of nanoparticles can be tailored by varying the most important synthesis conditions (described in Sections 3.1–3.7), including thermodynamic and non-thermodynamic processing variables (nature and stoichiometry of reactants, temperature, pH, pressure, process time, and additives).

3.1 Supersaturation, nucleation, and crystal growth

Supersaturation is the most critical crystallization parameter, and, at a given temperature, it is commonly related to a higher amount of a substance in solution than that required for saturation.

Long shape (L1 < L2)	**Size range**	**L2/L1 ratio**
Rod Needle Wire	L1 = 3 nm–300 μm L2 = 10 nm–800 μm	L2/L1 (rods) = 5–15 L2/L1 (needles) = 20–80 L2/L1 (wires) = 50–100
Whisker/fiber	L1 = 100–1000 nm L2 = 30–100 μm	L2/L1 = 100–300
Large shape (L1 > L2)	**Size range**	**L1/L2 ratio**
Sheet Plate	L1 = 40 nm–50 μm L2 = 5 nm–3 μm	L1/L2 = 10–20
Spherical shape	**Size range**	—
Organized rod sphere Organized sheet sphere	Ø = 500 nm–100 μm	—

Table 2.
Various HA nanoparticle shapes and achievable size range by hydrothermal processes (where L1 is the width and L2 is the high, (see **Figure 1***)).*

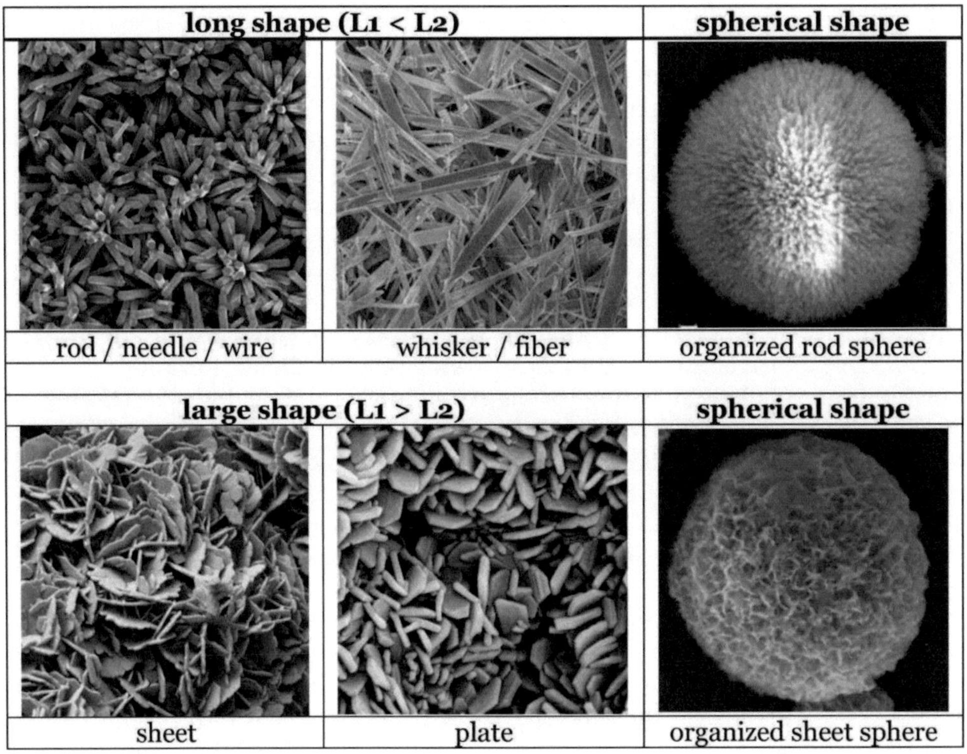

Figure 2.
Typical shapes of nanostructured HA.

It is the thermodynamic driving force that governs the formation mechanism of HA crystals, through nucleation and growth processes, and dictates the final crystal size distribution [44]. Nucleation and crystal growth are often competing mechanisms determining for the final crystal size and size distribution.

The thermodynamic driving force that induces the precipitation of HA under hydrothermal conditions is defined by the free energy variation in supersaturated solutions, which is given by the following relation in Eq. (1):

$$\Delta G = -(RT/n)\ln\left(A/K_{sp}\right) = -(RT/n)\ln(S) \qquad (1)$$

where ΔG is the molar Gibbs energy; R is the universal gas constant; T is the absolute temperature; n is the number of ions in HA molecule (for $Ca_5(PO_4)_3(OH)$ is n=5+3+1=9); and S is the supersaturation degree that is defined by the activity product of ion units (A) to the corresponding solubility product (K_{sp}) ratio, at a given temperature. Equation (1) also leads to the conclusion that the trend of absolute ΔG is directly related to the temperature. For example, if the temperature decreases, a slower precipitation rate of HA occurs and could lead to a larger crystal size.

The supersaturation degree (S) is defined as Eq. (2) where a_A, a_B, etc. are the activity of ionic species and K_{sp} is the solubility product constant. In the case of HA ($Ca_5(PO_4)_3(OH)$ or half unit cell), the supersaturation (S_{HA}) is given by Eq. (3), while its saturation index (SI) is defined as Eq. (4) [59]:

$$S = {a_A}^{a} \times {a_B}^{b} \times {a_C}^{c} \times / K_{sp} \qquad (2)$$

$$S_{HA} = \left(\left(a_{Ca}{}^{2+}\right)^{5} \times \left(a_{PO4}{}^{3-}\right)^{3} \times \left(a_{OH}{}^{-}\right)\right)/K_{spHA} \qquad (3)$$

$$SI_{HA} = \log(S_{HA}) \tag{4}$$

The control of supersaturation is critically prominent in hydrothermal condition to design HA crystals of the desired size, because it has a linear dependence with growth rate [72].

If SI < 0 the solution is undersaturated, and the mineral will dissolve into the solution; if SI = 0, the process dissolution and precipitation are in equilibrium.

When SI > 0 the solution is supersaturated, and the mineral will preferentially precipitate from the solution and grow. At positive and low SI, crystals can grow faster than their nucleation, resulting in a larger crystal formation; however, at higher SI crystal, nucleation prevails over crystal growth and smaller crystals are formed.

Some authors have shown that different morphologies ranging from rod-like, to plate-like to spherical HA nanoparticles could be obtained by varying the supersaturation values of the chemical reaction in hydrothermal conditions [31, 73]. SI value can also define a two-dimensional or layer-by-layer growth ($-\Delta G_{2D}$), in which two particles share a common crystallographic orientation and the growing interface can shift only transversely, leading the final structures to thin sheets and plate shaped. Alternatively, SI value can define a three-dimensional or continuous growth ($-\Delta G_{3D}$) where the interfaces can move without constraint [74].

3.2 Calcium- and phosphate-based precursors

HA is usually formed in the Ca–PO_4–H_2O hydrothermal system through a reaction between starting precursors, which contain Ca^{2+} (preferentially salts) and PO_4^{3-} ions, respectively. Each combination between Ca^{2+} and PO_4^{3-} sources can be used to create a specific method for preparing HA. Both ionic sources are mainly chosen as salts, to avoid the competitive formation of other CaP phases at external pH range of HA stability; for example, the use of phosphoric acid as phosphate source could precipitate the brushite phase [75, 76].

The most common chemical reactants used for the hydrothermal synthesis of HA are shown in **Table 3**, where the column of Ca^{2+} sources is split into soluble or insoluble forms.

Various synthetic strategies to form HA crystals were developed, and the related processes can be classified into two main types, based on the nature of Ca-based precursors that can be soluble or insoluble.

The mechanism of these processes follows several steps, which are schematically illustrated in **Figures 3** and **4**. In both cases, the hydrothermal treatment begins after the mixing of the precursors (and any additives).

By using the soluble calcium sources (**Figure 3**), the first step is the mixing of all soluble precursors in supersaturated conditions, followed by rapid nucleation, up to the formation of small nuclei (step of reaction between ions). The subsequent steps in hydrothermal conditions are the continuous growth of the nuclei and increasing of nanoparticle dimension by Ostwald ripening or maturation to reduce the overall energy (step of hydrothermal treatment) [73, 95].

Compared to synthesis with soluble salts, the alternative way (**Figure 4**) provides for an initial step of mixing of the precursors, by immerging the insoluble calcium salt in a phosphate solution. Then a dissolution-recrystallization mechanism proceeds at the solution/calcium sources interface in hydrothermal condition, through continuous nucleation-growth-ripening steps.

Calcium material undergoes a slow and gradual dissolution at the nanoscale, followed by the reaction between calcium and phosphate ions and by the nucleation/precipitation of HA crystals on the surface of dissolving calcium precursor. This process is thermodynamically favored due to the lower solubility product of

Ca^{2+} ***Soluble* sources**		PO_4^{3-} **Sources**
$Ca(CH_3COO)_2$	+	NaH_2PO_4 [61, 62];
$CaCl_2{\cdot}2H_2O$	+	K_2HPO_4 [48, 77]; NaH_2PO_4 [67]; Na_2HPO_4 [18, 34]; $(NH_4)_2HPO_4$ [78]; $C_8H_{10}NO_6P$ (PLP) [35]; $Na_5P_3O_{10}$ (STPP) [31]
$Ca(NO_3)_2{\cdot}4H_2O$	+	KH_2PO_4 [23]; K_2HPO_4 [47]; NaH_2PO_4 [58, 63]; Na_2HPO_4 [36, 64]; Na_3PO_4 [32, 37, 38]; $NH_4H_2PO_4$ [39, 65, 68]; $(NH_4)_2HPO_4$ [20, 29, 40–45, 49, 50, 54, 56, 59, 60, 66, 70, 79–83] $(NH_4)_2HPO_4$ + $NH_4H_2PO_4$ [71, 84]
Ca^{2+} ***Insoluble* sources**		PO_4^{3-} **Sources**
$Ca(OH)_2$	+	$(NH_4)_2HPO_4$ [85, 86] $CaHPO_4{\cdot}2H_2O$ [51]
$CaSO_4$	+	$(NH_4)_2HPO_4$ [87, 88]
$CaCO_3$	+	$(NH_4)_2HPO_4$ [89]; $CaHPO_4$ [46]
$CaCO_3$ (3-D)	+	$NH_4H_2PO_4$ [90]; $(NH_4)_2HPO_4$ [25, 91–93]; Na_3PO_4 [28]; K_2HPO_4 [94] $(NH_4)_2HPO_4$ + $NH_4H_2PO_4$ [26]
β-TCP	+	$CaHPO_4{\cdot}2H_2O$ (insolub.) [53]

Table 3.
Most popular precursors in the Ca–PO_4–H_2O hydrothermal system for HA production.

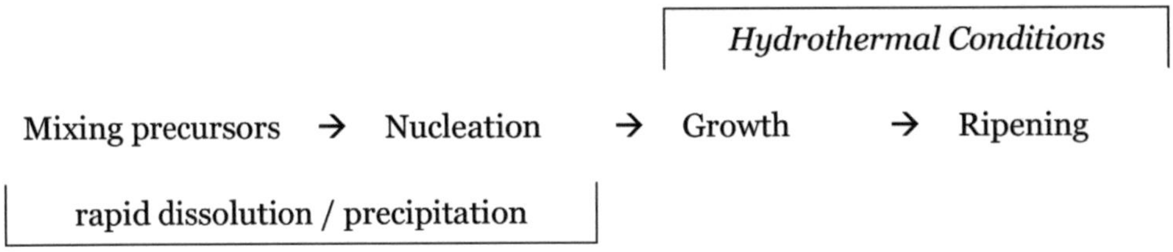

Figure 3.
Synthesis scheme by using soluble calcium-based precursor (powder form).

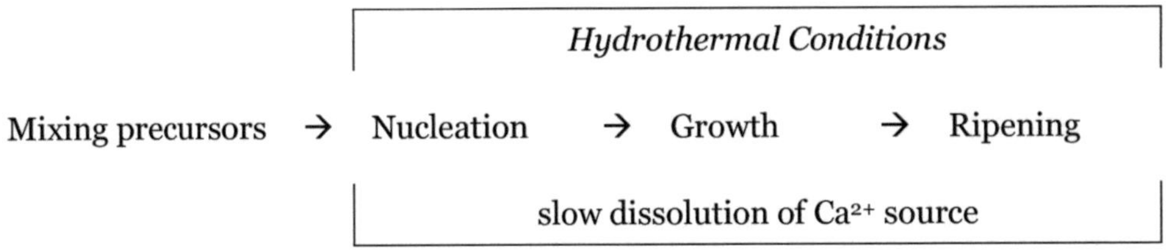

Figure 4.
Synthesis scheme by using insoluble calcium-based precursor (powder or 3-D block forms).

HA compared to calcium sources ($K_{psHA} << K_{psCaCO3} < K_{psCaSO4} < K_{psCa(OH)2}$). The solution where calcium source is soaked is initially undersaturated with respect to calcium; when source dissolves, it releases Ca^{2+}, and corresponding negative ions (e.g., CO_3^{2-} or OH^-) at the interface and a supersaturated condition, relating to HA, are created, and so it can start to precipitate and replace the original material.

The orientation of formed HA crystals may or may not have relationship with the parent calcium source crystal, with consequences on the structural control on the transformation from insoluble calcium source. But if calcium source has similar symmetry and unit cell dimensions of HA, the mineral replacement can occur by an epitactic or topotactic growth of HA crystals on the surface of dissolving precursor crystals, so that its orientation determines the disposition of HA nanocrystals. This mechanism is particularly evident during the conversion process of biogenic source made of calcite, where PO_4^{3-} ions replace CO_3^{2} at the interface to produce a 3-D structure of HA which accurately reproduces and preserves the whole complex morphology (external and internal) of the original calcite template [28]. Therefore, the greater versatility in tailoring of HA crystal size and morphology could be achieved through precipitation from homogeneous solutions containing both soluble calcium and phosphate sources. By adopting an insoluble salt of calcium as precursor, the degree of dissolution/precipitation, accompanied by interface reaction rate control, shall facilitate the formation of anisotropic particles which constitute the 3D structures.

3.3 Temperature

By shifting the dielectric constant and density with respect to temperature, water provides an excellent reaction medium for hydrothermal processing of HA nanostructured, leading to higher reaction rates and smaller particles than conventional processing route. According to the reaction temperature, hydrothermal synthesis can be classified into subcritical ($T < 374°C$) and supercritical ($T > 374°C$) synthesis, although these syntheses are typical subcritical, because they are easily applicable to industrial and laboratory operations. Subcritical conditions can be classified as mild (temperatures near to 100°C) or elevated (temperature up to 250°C). Many studies are done with synthesis temperature in the range from mild to elevated subcritical conditions (~100–240°C), showing the formation of various morphologies of particles with different sizes [18, 19, 21, 37, 39, 48, 53, 55, 60, 80, 86, 87, 89].

Hydrothermal processes carried out at work temperature higher than 250°C are rarely documented in literature; for this purpose, rapid and continuous hydrothermal synthesis could be considered an appropriate technique [41, 80]. This could lead to the deterioration of the internal vessel (frequently made of inert polymer) or potential additives used during the process. Furthermore, high pressure and specific volume of water in the proximity to the supercritical point could damage the hydrothermal device. Anyway, the advantage of syntheses at very high temperature was not yet clearly evidenced in literature.

The synthesis temperature significantly affects the precipitation/dissolution of HA, and it is a crucial crystal growth parameter, because it regulates the supersaturation by means of solubility products K_{ps}, that is temperature dependent. The following equation (Eq. (5)) can express the solubility product of HA as a function of temperature [21, 96]:

$$\log K_{spHA} = -(a/T + bT + c) \qquad (5)$$

where a, b, and c are constant.

As a general rule, high temperature leads to the formation of large and long fibers/particles, while at low temperature small dimensions are preferred [12, 46, 81]. This depend on predominant growth, according to specific kinetic of crystallization and colloidal stability of particles [72].

Increasing hydrothermal temperature affects the significantly improving phase purity and Ca/P ratio of HA precipitate [93]. Crystallinity features are also affected by the change of synthesis temperature; an increased crystallinity with increased temperature occurs compared with those prepared at a low temperature because nuclei growth tends to happen at higher temperatures [18, 47].

Du et al. described the hydrothermal process through two transition temperatures to particle growth can be at 160°C and a temperature between 200 and 220°C: the first transition is for uniformity, where particles synthesized at temperatures above it are more uniform, while the second limits the expansion of the particle length [97].

3.4 pH: phosphate equilibrium

pH is an important parameter that can be modified from alkaline to acidic condition to obtain the desired morphology of HA nanostructures.

pH condition performs the synthesis through two different ways: (i) by affecting the distribution of phosphate species in solution, according to the distribution of phosphate species (H_3PO_4, $H_2PO_4^-$, HPO_4^{2-}, and PO_4^{3-}) against the pH value, as well as (ii) the modification of the electrical charge on the HA crystal surface.

i. Different calcium phosphate intermediates are formed by adjusting the pH value; among them, the precipitation of HA may occur preferentially from slightly acidic condition up to basic pH conditions (approximately in the range of 5–11). Therefore, the solubility (or instability) of HA is highly affected under acidic conditions and decreases with increasing pH. The solubility isotherms of the CaP phases as a function of pH value and calcium (or phosphate) concentration from room temperature to hydrothermal conditions (e.g., T = 200°C) are shown in **Figure 5**. At 25°C the HA stability field exists at equilibrium pH > 4.8 (**Figure 5a**), whereas at 200°C the HA stability field extends to an equilibrium pH as low as 2.9 (**Figure 5b**).

An increment of pH value shifts the phosphate species equilibrium from $H_3PO_4 \rightarrow H_2PO_4^- \rightarrow HPO_4^{2-} \rightarrow PO_4^{3-}$, respectively, and increase the saturation index of HA according to Eq. (6) [74]:

$$\log SI_{HA} = 4\ \mathrm{pH} - 72.08 \tag{6}$$

This means that at pH~5 the SI = 1, while at higher pH the solution, is supersaturated and at lower pH is undersaturated. In other words, an increase in pH from about 5 results in increases in the nucleation rate of HA and its subsequent crystal growth.

ii. The pH affects the electric charge of the HA crystal surface by changing the distribution of hydroxyl groups or protons. It was experimentally estimated that the zero charge point for HA in solutions occurs close to neutral pH, for which, in basic or acidic conditions, the crystal surface will be negatively or positively charged, as result of preferred absorption of Ca^{2+} and PO_4^{3-} ions on the a-axis or c-axis; pH value results in anisotropic growth of the crystallites that will be into one-dimensional nanorods or two-dimensional nanoplates.

Alkaline conditions produce smaller and less elongated particles at any certain orientation, while the formation of rod-like or needle-like and c-axis elongated HA particles with hydrothermal process occurs at acidic conditions [20, 42, 43]. The variation in acidic pH has a pronounced influence on the length of the rod-like particles, due to the protonation trend of the a-axis.

More complex HA particles, including three-dimensional feathery, microfiber, and microcube structures were obtained by modulating pH during the synthesis [13].

In addition to temperature, the degree of crystallinity of final HA is also affected by difference in pH that can modify the nucleation and growth rates of the nuclei [42]; for example, higher crystallinity was observed in samples that were prepared at low pH conditions [8, 13].

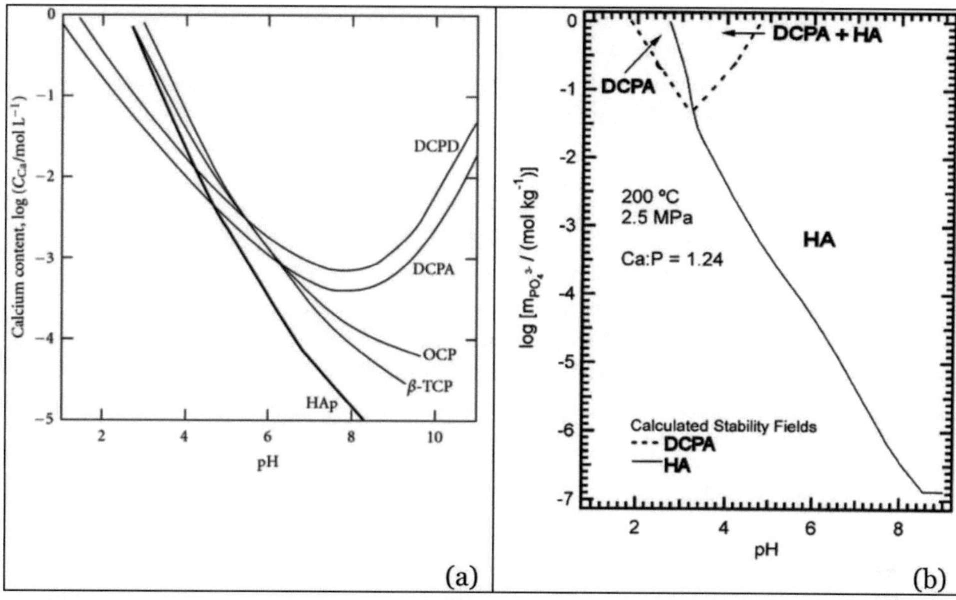

Figure 5.
Diagrams for the Ca–PO_4–H_2O system at 25°C (a) and 200°C (b) as a function of the equilibrium solution pH.

Sadat-Shojai et al. have recently reported that HA nanorods having high crystallinity and high aspect ratio could merely be prepared through precipitation at approximately neutral conditions followed by hydrothermal treatment at 200°C for 60 h. In this case, the nanoparticles show high dispersion stability, indicating their high surface charge and a low tendency for agglomeration [98].

The aspect ratio of HA particles is significantly affected by the pH of the reaction mixture; the mean aspect ratio of nanoparticles greatly decreases with increasing pH.

3.5 Process time

Process time, if prolonging, would enlarge the size of HA particles and also contribute to increasing the crystallinity. Its less significant factor affects the structural and morphological characteristics of HA nanoparticles, if compared with temperature and pH.

As temperature and pH affect the dissolution of HA, by maintaining these conditions as constants, it can be seen that this HA property is determined as a function of time. The concentration in hydrothermal solution of ionic species constituting HA is in relationship with process time, in accordance with classical parabolic behavior, expressed by Eq. (7),

$$[C]^2 = Kt + A \approx Kt \qquad (7)$$

where [C] is the concentration of ions, A is a constant, K is the parabolic rate constant, and t is the process time. For various hydrothermal temperature, a dissolution equilibrium time up to the achievement of plateau could be reached.

This equilibrium time of the reactions would be shortened with the increasing of hydrothermal temperature. For example, at neutral pH, Zhang et al. had observed an equilibrium time after 10 h at 300°C and after 12 h at 200°C [96].

Several studies report the linear increasing of the synthesized length of HA nanoparticles up to a synthesis time limit. At 150 and 200°C, respectively, Jin et al. and Earl et al. observed the elongation of nanorods within 24 h; for the range of

24–72 hours of synthesis, crystallinity and size are not significantly affected by the differences in synthesis duration [20, 37].

Over the equilibrium time, self-assembly phenomena can occur toward more complex forms, i.e., flower-like morphologies [68], while in 3-D biomorphic conversion, prolonged heating time could influence the aggregation and morphology of HA through a rearrangement of the building blocks such as nanosheets and nanorods.

Hydrothermal treatment can be done in various time-temperature profiles designed as serious, stepwise, slow, intervallic thermal heating, etc. By carefully changing these profiles, the final morphologies of the resulting powder can be plates, hexagonal prisms, needles, and fine-plates, respectively [29].

The prolonged time can affect the colloidal stability and decomposition of any additives that will further be discussed in Section 3.7.

3.6 Concentration/ion ratio

The final HA morphologies are sensitive to the concentration of ions during the reaction, defining the degree of supersaturation (Eqs. (2) and (3)).

Depending only by the concentration of the precursors, the HA nanostructures can form different morphologies. Hao et al. have observed that HA particles could have irregular or uniform plate-like forms, which tend to aggregate into small fibrous bundles, by increasing calcium concentration in solution [59].

Some authors claim that the diameter of the particles increases with the increment of reactant concentration because less concentrated chemical material limits the ion transportation. The increments of the mean length of the particles are also concluded to be significantly affected by the concentration of the initial material [42].

It is known that HA powders can be prepared with controlled Ca/P molar ratio (stoichiometric value is 1.67) in accordance to the starting amount of Ca^{2+} and PO_4^{3-} in solution [29, 98]. If the starting calcium precursor is $CaCO_3$, the resulting HA products will be not stoichiometric, due to partial substitutions of the anions (PO_4^{3-}, and OH^-) by CO_3^{2-} groups.

3.7 Additives

Recently, it was found that additional components are usually included with the precursors to modify the morphology and size of the produced HA nanostructures, during hydrothermal synthesis. Some of the most popular additives used in many hydrothermal syntheses are listed in **Table 4** and classified on the basis of their chemical characteristic. They include organic or inorganic matrices.

The intrinsically different surface reactivity of HA crystals (positively charged Ca^{2+}-rich c-surface and negatively charged OH^-- and PO_4^{3-}-rich a-surface) can selectively adsorb various ions, organic compounds, and proteins.

Long HA whiskers can be prepared by homogeneous hydrothermal precipitation using acetamide, as a result of hydrolysis at elevated temperature and following interaction between acetic ions with their negative charges and the a-sites of HA [60]. This match also characterizes citrate salts [38].

Urea influences the increase of the diameter of HA nanoparticles because its hydrolysis introduces CO_3^{2-} ions, incorporated inside the HA crystal structure; therefore, it controls the pH value through NH_3 release, affecting the nucleation and growth of HA crystals. Intensive decomposition of urea increases supersaturation and subsequently produces more nuclei and smaller crystals [42].

Chelating agent (e.g. EDTA or tartrate) can induce the formation of a ring-like structure through the complex with Ca^{2+} contained in HA. Final particles can be obtained after decomposition of these structures in controlled hydrothermal treatment.

Properties	Additive
Electrolytes	Acetamide (AA) [60], urea [29, 42, 58, 68], Na-citrate [37, 38, 44], propionamide (PA) [59], etc.
Chelating agents	Ethylenediaminetetraacetic acid (EDTA) [36], K-Na-tartrate [67], etc.
Organic modifiers/ surfactants	Cetyltrimethylammonium bromide (CTAB) [77, 99], glutamic acid (Glu) [31, 66], liposome [83], polyethylene glycol (PEG) [43], polyvinylpyrrolidone (PVP) [39], macromolecule (HTCC) [33], etc.
Proteins	Hemoglobin [62], lysine (Ly) [56], etc.

Table 4.
Various additives commonly used during hydrothermal synthesis of calcium phosphate.

As a result of complexation, the concentration of free calcium ions dramatically decreases, leading to the HA nuclei with smaller size and quantity. During the hydrothermal treatment, calcium ions are released, and each nucleus will grow to the distinct single needle-like particles and finally to well-separated long fibers. It is well known that increasing pH value has a considerable improvement on complexes stability. The subsequent hydrothermal treatment leads to the Ca-complex decomposition and HA crystals are grown on the previously formed pattern, according to anisotropic growth along the c-axis.

Surfactants are usually used as the controlling reagent to prepare 3-D architecture of HA-based materials. Contrary to anionic reactant like EDTA, the cationic surfactant CTAB can bind with PO_4^{3-} of reaction system by the ionic charge and stereochemistry due to their complementarity tetrahedral structures, so that phosphate anion can be incorporated to the formed nuclei, and the final HA and size of HA particles can be well controlled. During the hydrothermal process, CTAB-HA complexes are created, and their controlled coalescence produce nanorods with uniform morphology and controllable size [99].

Some attempts have been made to exploit the developed organic materials in controlling the crystal growth of HA. Zhu et al. [33] synthesized rod-like HA nanoparticles of various aspect ratios by means of the hydrothermal method in the presence of HTCC as a cationic polymer template. HTCC molecules are first incorporated into PO_4^{3-} and OH^- anions by charge and stereochemical complementarily, and the rigid chains of polymer molecules are then converted to extended chains, connected through the ion bonds of PO_4^{3-} to form a 2D structure, followed by formation of a 3D rod-like morphology through self-assembling process via hydrogen bond interactions. Nucleation and subsequent crystal growth can occur upon adding the Ca precursor and hydrothermal treatment, respectively.

Recently, a novel hydrothermal method based on the liquid-solid-solution strategy has been developed to synthesize surface-modified HA nanorods of various aspect ratios. According to this strategy, controlled growth of HA nanorods with tunable morphology can be achieved by adequately tuning the interfaces between surfactants and the central atoms of HA. Nanostructured HA with desired characteristics can be fabricated by using organic modifiers, such as PEG, Tween-20, sorbitol, etc., in various hydrothermal conditions. For example, they favor the formation of HA nanorods with a larger aspect ratio, or small-sized, or long length, respectively, at high or low synthesis temperature.

Some proteins have an affinity to a specific face of HA crystals, and they are strong inhibitors for perpendicular growth to the adsorbed face, in hydrothermal conditions at low temperature. For example, if the c-plane is covered by proteins (e.g., hemoglobin molecules), the crystal will preferentially grow in the direction of the a-axis to form HA nanosheets. Then the hemoglobin could adsorb on the surface of newly formed HA nanosheets to increase their dimension. At high temperature, the hemoglobin could be hydrolyzed and decomposed under the hydrothermal conditions, and the degraded products could be differently adsorbed by the formed HA nanosheets up to the formation of different structures (e.g., flower-like or dandelion-like morphologies) [62].

4. Conclusions

The hydrothermal method is particularly appropriate for the realization of good-quality crystals and allows tailored HA nanosized particles to be synthesized or characteristic of 3-D structures.

This chapter discussed about the most influencing variables in the hydrothermal synthesis of HA and their effects to prepare a wide range of crystal morphologies.

Among the considered variables, temperature and pH seem to be the most significant factors affecting the dimensional, geometry, and crystalline characteristics of HA nanoparticles, e.g., through face-selective interaction and anisotropic growth processes.

Table 5 summarizes the process parameters discussed and their effects and contribution on the final structures, which can have the form of hexagonal prisms (rods, needles, wires), sheets, plates, whiskers, fine spherical-shaped, etc., of varying sizes.

This extensive range of morphologies may be useful for developing various and improved types of nanostructured biomaterials, which could have potential application for repairing bone defects and tissue regeneration.

Conditions	Effects	Low values	High values
Supersaturation	Crystal size, size distribution	Large particles, 2D growth	Small particles, 3D growth
Selection of precursors	HA precipitation, crystals orientation	—	—
Concentration	Saturation, Ca/P molar ratio control	—	Increase particle dimension
Temperature	Solubility, nucleation, growth	Small particles, low crystallinity	Large particles, high phase purity, high crystallinity
pH	Crystal growth rate, anisotropic growth of crystals	Small particles, preferential growth along c-axis (rod shape), high crystallinity	Large particles, preferential growth along a-axis (plate shape)
Process time	Ripening	Small size	Large size, high crystallinity
Additives	Structures control, complex geometry	Various effects	Various effects

Table 5.
Summary scheme of the hydrothermal parameters contribution on the final HA nanostructure.

Conflict of interest

The authors declare no conflict of interest.

Author details

Andrea Ruffini*, Simone Sprio, Lorenzo Preti and Anna Tampieri
CNR-ISTEC – Institute of Science and Technology for Ceramics, Faenza, RA, Italy

*Address all correspondence to: andrea.ruffini@istec.cnr.it

References

[1] Bohner M. Calcium orthophosphates in medicine: From ceramics to calcium phosphate cements. Injury. 2000;**31**:D37-D47

[2] Sprio S, Sandri M, Ruffini A, Adamiano A, Iafisco M, Dapporto M, et al. Tissue engineering and biomimetics with bioceramics. In: Advances in Ceramic Biomaterials. Elsevier, Ltd., Woodhead Publishing; 2017. pp. 407-432

[3] He W, Andersson M. Biomimetic synthesis of nanostructured calcium phosphates. In: Frontiers in Nanobiomedical Research. Singapore: The World Scientific Encyclopedia of Nanomedicine and Bioengineering II. 2017. pp. 85-124

[4] Kumta PN, Sfeir C, Lee D-H, Olton D, Choi D. Nanostructured calcium phosphates for biomedical applications: Novel synthesis and characterization. Acta Biomaterialia. 2005;**1**(1):65-83

[5] Hui P, Meena SL, Singh G, Agarawal RD, Prakash S. Synthesis of hydroxyapatite bio-ceramic powder by hydrothermal method. Journal of Minerals and Materials Characterization and Engineering. 2010;**09**(08):683-692

[6] Weiner S, Traub W, Wagner HD. Lamellar bone: Structure–function relations. Journal of Structural Biology. 1999;**126**(3):241-255

[7] Lee J-W, Yun H-S, Nakano T. Induction of biological apatite orientation as a bone quality parameter in bone regeneration using hydroxyapatite/poly caprolactone composite scaffolds. Tissue Engineering. Part C, Methods. 2016;**22**(9):856-863

[8] Liu Q, Huang S, Matinlinna JP, Chen Z, Pan H. Insight into biological apatite: Physiochemical properties and preparation approaches. BioMed Research International. 2013;**2013**:1-13

[9] Ferraz MP, Monteiro FJ, Manuel CM. Hydroxyapatite nanoparticles: A review of preparation methodologies. Journal of Applied Biomaterials & Biomechanics. 2004;**2**(2):74-80

[10] Fatimah M, Shaaban ASA, Seliman SSS. Overview: Process parameters for hydrothermal synthesis of hydroxyapatite. Journal of Advanced Manufacturing Technology. 2012;**6**(1):47-59

[11] Chen F, Zhu Y, Wu J, Huang P, Cui D. Nanostructured calcium phosphates: Preparation and their application in biomedicine. Nano Biomedicine and Engineering. 2012;**4**(1):41-49

[12] Loo SCJ, Siew YE, Ho S, Boey FYC, Ma J. Synthesis and hydrothermal treatment of nanostructured hydroxyapatite of controllable sizes. Journal of Materials Science. Materials in Medicine. 2008;**19**(3):1389-1397

[13] Sadat-Shojai M, Khorasani M-T, Dinpanah-Khoshdargi E, Jamshidi A. Synthesis methods for nanosized hydroxyapatite with diverse structures. Acta Biomaterialia. 2013;**9**(8):7591-7621

[14] Dorozhkin SV, Epple M. Biological and medical significance of calcium phosphates. Angewandte Chemie, International Edition. 2002;**41**(17):3130-3146

[15] Wang L, Nancollas GH. Calcium orthophosphates: Crystallization and dissolution. Chemical Reviews. 2008;**108**(11):4628-4669

[16] Okada M, Matsumoto T. Synthesis and modification of apatite nanoparticles for use in dental and medical applications. Japanese Dental Science Review. 2015;**51**(4):85-95

[17] Iwamoto T, Hieda Y, Kogai Y. Effect of hydroxyapatite surface morphology

on cell adhesion. Materials Science and Engineering: C. 2016;**69**:1263-1267

[18] Li R, Cai Y, Yang Y, Fang S, Su X, Sun Y, et al. Dimensionally and morphologically controlled growth of calcium phosphate crystals by an organic-free hydrothermal process. Ceramics International. 2016;**42**(15):17387-17397

[19] Mohan N, Palangadan R, Varma H. Hydroxyapatite scaffolds constituting highly oriented crystals derived from synthetic precursors by hydrothermal reactions. Ceramics International. 2016;**42**(15): 17259-17268

[20] Earl JS, Wood DJ, Milne SJ. Hydrothermal synthesis of hydroxyapatite. Journal of Physics Conference Series. 2006;**26**:268-271

[21] Byrappa K, Adschiri T. Hydrothermal technology for nanotechnology. Progress in Crystal Growth and Characterization of Materials. 2007;**53**(2):117-166

[22] Yoshimura M, Sujaridworakun P, Koh F, Fujiwara T, Pongkao D, Ahniyaz A. Hydrothermal conversion of calcite crystals to hydroxyapatite. Materials Science and Engineering: C. 2004;**24**(4):521-525

[23] Ashok M, Kalkura SN, Sundaram NM, Arivuoli D. Growth and characterization of hydroxyapatite crystals by hydrothermal method. Journal of Materials Science: Materials in Medicine. 2007;**18**(5):895-898

[24] Hiroaki O, Shin Y. Hydrothermal synthesis of calcium phosphate from Corbicula shells. Journal of Ecotechnology Research. 2016;**18**(1):7-10

[25] Vecchio KS, Zhang X, Massie JB, Wang M, Kim CW. Conversion of bulk seashells to biocompatible hydroxyapatite for bone implants. Acta Biomaterialia. 2007;**3**(6):910-918

[26] Ruffini A, Sprio S, Tampieri A. Study of the hydrothermal transformation of wood-derived calcium carbonate into 3D hierarchically organized hydroxyapatite. Chemical Engineering Journal. 2013;**217**:150-158

[27] Jinawath S, Polchai D, Yoshimura M. Low-temperature, hydrothermal transformation of aragonite to hydroxyapatite. Materials Science and Engineering: C. 2002;**22**(1):35-39

[28] Álvarez-Lloret P, Rodríguez-Navarro AB, Falini G, Fermani S, Ortega-Huertas M. Crystallographic control of the hydrothermal conversion of calcitic Sea urchin spine (paracentrotus lividus) into apatite. Crystal Growth & Design. 2010;**10**(12):5227-5232

[29] Neira IS, Kolen'ko YV, Lebedev OI, Van Tendeloo G, Gupta HS, Guitián F, et al. An effective morphology control of hydroxyapatite crystals via hydrothermal synthesis. Crystal Growth & Design. 2009;**9**(1):466-474

[30] Jinawath S, Pongkao D, Suchanek W, Yoshimura M. Hydrothermal synthesis of monetite and hydroxyapatite from monocalcium phosphate monohydrate. International Journal of Inorganic Materials. 2001;**3**(7):997-1001

[31] Yang Y, Wu Q, Wang M, Long J, Mao Z, Chen X. Hydrothermal synthesis of hydroxyapatite with different morphologies: Influence of supersaturation of the reaction system. Crystal Growth & Design. 2014;**14**(9):4864-4871

[32] Su J, Liang, Yunfei, Shuhua, Song, Zhuojing. Effects of hydrothermal treatment on the properties of

nanoapatite crystals. International Journal of Nanomedicine. 2012;**7**:5151-5158

[33] Zhu A, Lu Y, Si Y, Dai S. Fabricating hydroxyapatite nanorods using a biomacromolecule template. Applied Surface Science. 2011;**257**(8):3174-3179

[34] Lim H, Kassim A, Ming H. Preparation and characterization of calcium phosphate nanorods using reverse microemulsion and hydrothermal processing routes. Sains Malaysiana. 2010;**39**:267-273

[35] Zhao X-Y, Zhu Y-J, Lu B-Q, Chen F, Qi C, Zhao J, et al. Hydrothermal synthesis of hydroxyapatite nanorods using pyridoxal-5′-phosphate as a phosphorus source. Materials Research Bulletin. 2014;**55**:67-70

[36] Xin R, Ren F, Leng Y. Synthesis and characterization of nano-crystalline calcium phosphates with EDTA-assisted hydrothermal method. Materials and Design. 2010;**31**(4):1691-1694

[37] Jin X, Chen X, Cheng Y, Wang L, Hu B, Tan J. Effects of hydrothermal temperature and time on hydrothermal synthesis of colloidal hydroxyapatite nanorods in the presence of sodium citrate. Journal of Colloid and Interface Science. 2015;**450**:151-158

[38] Jin X, Zhuang J, Zhang Z, Guo H, Tan J. Hydrothermal synthesis of hydroxyapatite nanorods in the presence of sodium citrate and its aqueous colloidal stability evaluation in neutral pH. Journal of Colloid and Interface Science. 2015;**443**:125-130

[39] Hu J, Zhou Y, Huang L, Liu J, Lu H. Effect of nano-hydroxyapatite coating on the osteoinductivity of porous biphasic calcium phosphate ceramics. BMC Musculoskeletal Disorders. 2014;**15**(1):1-11

[40] Buitrago-Vásquez M, Ossa-Orozco CP. Hydrothermal synthesis of hydroxyapatite nanorods using a fruit extract template. DYNA. 2018;**85**(204):283-288

[41] Chaudhry AA, Haque S, Kellici S, Boldrin P, Rehman I, Khalid FA, et al. Instant nano-hydroxyapatite: A continuous and rapid hydrothermal synthesis. Chemical Communications. 2006;**21**:2286-2288

[42] Sadat-Shojai M, Atai M, Nodehi A. Design of experiments (DOE) for the optimization of hydrothermal synthesis of hydroxyapatite nanoparticles. Journal of the Brazilian Chemical Society. 2011;**22**(3):571-582

[43] Salariana M, Solati-Hashjin M, Goudarzi A, Shafiei SS, Salarian R, Nemati ZA. Effect of surfactant in formation of hydroxyapatite Nano-rods under hydrothermal conditions. Iranian Journal of Pharmaceutical Sciences. 2008;**4**(2):157-162

[44] Yang H, Zeng H, Hao L, Zhao N, Du C, Liao H, et al. Effects of hydroxyapatite microparticle morphology on bone mesenchymal stem cell behavior. Journal of Materials Chemistry B. 2014;**2**(29):4703-4710

[45] Yubao L, De Wijn J, Klein CPAT, Van De Meer S, De Groot K. Preparation and characterization of nanograde osteoapatite-like rod crystals. Journal of Materials Science. Materials in Medicine. 1994;**5**(5):252-255

[46] Zhang X, Vecchio KS. Hydrothermal synthesis of hydroxyapatite rods. Journal of Crystal Growth. 2007;**308**(1):133-140

[47] Méndez-Lozano N, Velázquez-Castillo R, Rivera-Muñoz EM, Bucio-Galindo L, Mondragón-Galicia G, Manzano-Ramírez A, et al. Crystal growth and structural analysis of hydroxyapatite nanofibers

synthesized by the hydrothermal microwave-assisted method. Ceramics International. 2017;**43**(1):451-457

[48] Ban S, Hasegawa J. Morphological regulation and crystal growth of hydrothermal-electrochemically deposited apatite. Biomaterials. 2002;**23**(14):2965-2972

[49] Chen C, Huang Z, Yuan W, Li J, Cheng X, Chi R. Pressure effecting on morphology of hydroxyapatite crystals in homogeneous system. CrystEngComm. 2011;**13**(5):1632-1637

[50] Yubao L, De Groot K, De Wijn J, Klein CPAT, Meer SVD. Morphology and composition of nanograde calcium phosphate needle-like crystals formed by simple hydrothermal treatment. Journal of Materials Science. Materials in Medicine. 1994;**5**(6-7):326-331

[51] Goto T, Kamitakahara M, Kim IY, Ohtsuki C. Effects of ethanol addition on formation of hydroxyapatite through hydrothermal treatment of dicalcium phosphate dihydrate. IOP Conference Series: Materials Science and Engineering. 2011;**18**(19):192015

[52] Han Y, Xu K, Montay G, Fu T, Lu J. Evaluation of nanostructured carbonated hydroxyapatite coatings formed by a hybrid process of plasma spraying and hydrothermal synthesis. Journal of Biomedical Materials Research. 2002;**60**(4):511-516

[53] Watanabe T, Kawachi G, Kamitakahara M, Kikuta K, Ohtsuki C. Formation of needle-like hydroxyapatite by hydrothermal treatment of CaHPO4·2H2O combined with β-Ca3(PO4)2. Journal of the Ceramic Society of Japan. 2009;**117**(1366):759-764

[54] Xiao D, Guo T, Yang F, Feng G, Shi F, Li J, et al. In situ formation of nanostructured calcium phosphate coatings on porous hydroxyapatite scaffolds using a hydrothermal method and the effect on mesenchymal stem cell behavior. Ceramics International. 2017;**43**(1):1588-1596

[55] Lin K, Chang J, Liu X, Chen L, Zhou Y. Synthesis of element-substituted hydroxyapatite with controllable morphology and chemical composition using calcium silicate as precursor. CrystEngComm. 2011;**13**(15):4850

[56] Chen YQ, Xing XF, Gao WM. Synthesis of spherical nano-hydroxyapatite by hydrothermal method with L-lysine template. Key Engineering Materials. 2014;**633**:17-20

[57] Costa DO, Dixon SJ, Rizkalla AS. One- and three-dimensional growth of hydroxyapatite nanowires during sol–gel–hydrothermal synthesis. ACS Applied Materials & Interfaces. 2012;**4**(3):1490-1499

[58] Jokić B, Mitrić M, Radmilović V, Drmanić S, Petrović R, Janaćković D. Synthesis and characterization of monetite and hydroxyapatite whiskers obtained by a hydrothermal method. Ceramics International. 2011;**37**(1):167-173

[59] Hao L, Yang H, Zhao N, Du C, Wang Y. Controlled growth of hydroxyapatite fibers precipitated by propionamide through hydrothermal synthesis. Powder Technology. 2014;**253**:172-177

[60] Zhang H, Darvell BW. Formation of hydroxyapatite whiskers by hydrothermal homogeneous precipitation using acetamide: Formation of hydroxyapatite whiskers. Journal of the American Ceramic Society. 2011;**94**(7):2007-2013

[61] Zhao X-Y, Zhu Y-J, Chen F, Lu B-Q, Wu J. Nanosheet-assembled hierarchical nanostructures of hydroxyapatite: Surfactant-free microwave-hydrothermal rapid synthesis, protein/DNA adsorption and pH-controlled release. CrystEngComm. 2013;**15**(1):206-212

[62] Zhao X-Y, Zhu Y-J, Zhao J, Lu B-Q, Chen F, Qi C, et al. Hydroxyapatite nanosheet-assembled microspheres: Hemoglobin-templated synthesis and adsorption for heavy metal ions. Journal of Colloid and Interface Science. 2014;**416**:11-18

[63] Guan J-J, Tian B, Tang S, Ke Q-F, Zhang C-Q, Zhu Z-A, et al. Hydroxyapatite coatings with oriented nanoplate arrays: Synthesis, formation mechanism and cytocompatibility. Journal of Materials Chemistry B. 2015;**3**(8):1655-1666

[64] Zhang H, Zhou K, Li Z, Huang S. Plate-like hydroxyapatite nanoparticles synthesized by the hydrothermal method. Journal of Physics and Chemistry of Solids. 2009;**70**(1):243-248

[65] Lin K, Zhou Y, Zhou Y, Qu H, Chen F, Zhu Y, et al. Biomimetic hydroxyapatite porous microspheres with co-substituted essential trace elements: Surfactant-free hydrothermal synthesis, enhanced degradation and drug release. Journal of Materials Chemistry. 2011;**21**(41):16558

[66] Qi M, Qi J, Xiao G, Zhang K, Lu C, Lu Y. One-step hydrothermal synthesis of carbonated hydroxyapatite porous microspheres with a large and uniform size regulated by l-glutamic acid. CrystEngComm. 2016;**18**(31):5876-5884

[67] Ma M-G. Hierarchically nanostructured hydroxyapatite: Hydrothermal synthesis, morphology control, growth mechanism, and biological activity. International Journal of Nanomedicine. 2012;7:1781-1791

[68] Lin K, Chang J, Zhu Y, Wu W, Cheng G, Zeng Y, et al. A facile one-step surfactant-free and low-temperature hydrothermal method to prepare uniform 3D structured carbonated apatite flowers. Crystal Growth & Design. 2009;**9**(1):177-181

[69] Aldrich DS, Smith MA. Pharmaceutical applications of infrared microspectroscopy. Applied Spectroscopy Reviews. 1999;**34**(4):275-327

[70] Neira IS, Guitián F, Taniguchi T, Watanabe T, Yoshimura M. Hydrothermal synthesis of hydroxyapatite whiskers with sharp faceted hexagonal morphology. Journal of Materials Science. 2008;**43**(7):2171-2178

[71] Riman R. Solution synthesis of hydroxyapatite designer particulates. Solid State Ionics. 2002;**151**(1-4):393-402

[72] Koutsopoulos S. Kinetic study on the crystal growth of hydroxyapatite. Langmuir. 2001;**17**(26):8092-8097

[73] Pang YX, Bao X. Influence of temperature, ripening time and calcination on the morphology and crystallinity of hydroxyapatite nanoparticles. Journal of the European Ceramic Society. 2003;**23**(10):1697-1704

[74] Viswanath B, Ravishankar N. Controlled synthesis of plate-shaped hydroxyapatite and implications for the morphology of the apatite phase in bone. Biomaterials. 2008;**29**(36):4855-4863

[75] Onoda H, Yamazaki S. Homogenous hydrothermal synthesis of calcium phosphate with calcium carbonate and corbicula shells. Journal of Asian Ceramic Societies. 2016;**4**(4):403-406

[76] Prakash KH, Kumar R, Ooi CP, Cheang P, Khor KA. Apparent solubility of hydroxyapatite in aqueous medium and its influence on the morphology of nanocrystallites with precipitation temperature. Langmuir. 2006;**22**(26):11002-11008

[77] Wang Y, Zhang S, Wei K, Zhao N, Chen J, Wang X. Hydrothermal synthesis of hydroxyapatite

nanopowders using cationic surfactant as a template. Materials Letters. 2006;**60**(12):1484-1487

[78] Manoj M, Subbiah R, Mangalaraj D, Ponpandian N, Viswanathan C, Park K. Influence of growth parameters on the formation of hydroxyapatite (HAp) nanostructures and their cell viability studies. Nano. 2015;**2**:2

[79] Rafique MMA. Hydrothermal processing of phase pure and doped hydroxyapatite and its characterization. Journal of Encapsulation and Adsorption Sciences. 2018;**08**(01):18-37

[80] Chaudhry AA, Knowles JC, Rehman I, Darr JA. Rapid hydrothermal flow synthesis and characterisation of carbonate- and silicate-substituted calcium phosphates. Journal of Biomaterials Applications. 2013;**28**(3):448-461

[81] Jiang D, Zhang J. Calcium phosphate with well controlled nanostructure for tissue engineering. Current Applied Physics. 2009;**9**(3):S252-S256

[82] Nagata F, Yamauchi Y, Tomita M, Kato K. Hydrothermal synthesis of hydroxyapatite nanoparticles and their protein adsorption behavior. Journal of the Ceramic Society of Japan. 2013;**121**(1417):797-801

[83] Ramedani A, Yazdanpanah A, Moztarzadeh F, Mozafari M. On the use of nanoliposomes as soft templates for controlled nucleation and growth of hydroxyapatite nanocrystals under hydrothermal conditions. Ceramics International. 2014;**40**(7):9377-9381

[84] Vecbiskena L, de Nardo L, Chiesa R. Nanostructured calcium phosphates for biomedical applications. Key Engineering Materials. 2014;**604**:212-215

[85] Jokanović V, Izvonar D, Dramićanin MD, Jokanović B, Živojinović V, Marković D, et al. Hydrothermal synthesis and nanostructure of carbonated calcium hydroxyapatite. Journal of Materials Science: Materials in Medicine. 2006;**17**(6):539-546

[86] Ortiz SL, Avila JH, Gutierrez MP, Gómez-Pozos H, Karthik TVK, Lugo VR. Hydrothermal synthesis and characterization of hydroxyapatite microstructures. In: 14th International Conference on Electrical Engineering, Computing Science and Automatic Control (CCE). IEEE Xplore Digital online. 2017. pp. 1-4

[87] Bingöl OR, Durucan C. Hydrothermal synthesis of hydroxyapatite from calcium Sulfate hemihydrate. American Journal of Biomedical Sciences. 2012;**4**(1):50-59

[88] Katsuki H, Furuta S, Komarneni S. Microwave- versus conventional-hydrothermal synthesis of hydroxyapatite crystals from gypsum. Journal of the American Ceramic Society. 2004;**82**(8):2257-2259

[89] Wu Y-S, Chang W-K, Jou M. Hydroxyapatite synthesised from nanosized calcium carbonate via hydrothermal method. Materials and Technologies. 2012;**27**(1):119-123

[90] Macha I, Boonyang U, Cazalbou S, Ben-Nissan B, Charvillat C, Oktar F, et al. Comparative study of coral conversion, part 2: Microstructural evolution of calcium phosphate. Journal of The Australian Ceramic Society. 2015;**51**:149

[91] Felício-Fernandes G, Laranjeira MCM. Calcium phosphate biomaterials from marine algae. Hydrothermal synthesis and characterisation. Química Nova. 2000;**23**(4):441-446

[92] Macha IJ, Grossin D, Ben-Nissan B. Conversion of marine structures to calcium phosphate materials: Mechanisms of conversion using two

different phosphate solutions. Key Engineering Materials. 2016;**696**:36-39

[93] Xu Y, Wang D, Yang L, Tang H. Hydrothermal conversion of coral into hydroxyapatite. Materials Characterization. 2001;**47**(2):83-87

[94] Tampieri A, Sprio S, Ruffini A, Celotti G, Lesci IG, Roveri N. From wood to bone: Multi-step process to convert wood hierarchical structures into biomimetic hydroxyapatite scaffolds for bone tissue engineering. Journal of Materials Chemistry. 2009;**19**(28):4973

[95] Kuśnieruk S, Wojnarowicz J, Chodara A, Chudoba T, Gierlotka S, Lojkowski W. Influence of hydrothermal synthesis parameters on the properties of hydroxyapatite nanoparticles. Beilstein Journal of Nanotechnology. 2016;7:1586-1601

[96] Zhang H, Li S, Yan Y. Dissolution behavior of hydroxyapatite powder in hydrothermal solution. Ceramics International. 2001;**27**(4):451-454

[97] Du X, Chu Y, Xing S, Dong L. Hydrothermal synthesis of calcium hydroxyapatite nanorods in the presence of PVP. Journal of Materials Science. 2009;**44**(23):6273-6279

[98] Sadat-Shojai M, Atai M, Nodehi A, Khanlar LN. Hydroxyapatite nanorods as novel fillers for improving the properties of dental adhesives: Synthesis and application. Dental Materials. 2010;**26**(5):471-482

[99] Wang Y, Chen J, Wei K, Zhang S, Wang X. Surfactant-assisted synthesis of hydroxyapatite particles. Materials Letters. 2006;**60**(27):3227-3231

Chapter 2

Dentin Materials as Biological Scaffolds for Tissue Engineering

Masaru Murata, Naoto Okubo, Mamata Shakya, Md Arafat Kabir, Kenji Yokozeki, Bowen Zhu, Masahiro Ishikawa, Ryuji Kitamura and Toshiyuki Akazawa

Abstract

Vital tooth-derived demineralized dentin matrix (DDM) has a bone-inductive ability, while non-vital tooth-derived DDM lost it. Acid treatment for dentin provides the increase of surface area, the release of matrix-binding growth factors such as BMPs, and the decrease of the infection risk. Human autograft of vital tooth-derived DDM was achieved first in Japan 2002, while first bone autograft was noted in Italy 1820. This paper introduced dentin/bone biology and a unique clinical case, combined with two types of non-vital tooth-derived DDM (roots, granules) for lateral bone augmentation. A 63-year-old woman revealed highly atrophic mandible in 2015. Three non-vital teeth were extracted, changed in shape, demineralized in 2% HNO_3, were rinsed, and were grafted immediately. The CT images at 3 months after the graft showed remarkable lateral augmentation. DDM scaffolds were received to host, and two fixtures were placed into the DDM-augmented bone. The patient was successfully restored with their own DDM scaffolds and implant surgery.

Keywords: dentin, bone, material, scaffold

1. Introduction

Regenerative medicine is based on applied biomaterial science. Biomaterials have a strong impact on the patient cure for improving the quality of life. Recently, absorbable materials have been needed in the field of tissue engineering. We have been challenging to develop bioabsorbable dentin scaffolds [1–13], harmonized with bone remodeling, by using the ultrasonic acid-etching technology [14–17] (**Figure 1**).

2. History of bone graft

First bone graft was allograft from dog to human in 1682. While xeno-or allograft of bone was a major in the Western countries during the Middle Ages, human bone autograft was noted first in 1820, Italy. The use of bovine-demineralized bone matrix (DBM) was reported first in 1889. The xenogenic DBM was grafted into human skeletal defects caused by osteomyelitis. Many DBM papers were published in the twentieth century [18]. Generally, fresh autogenous bone remains the gold standard of treatment for nonunion and large bone defects [19].

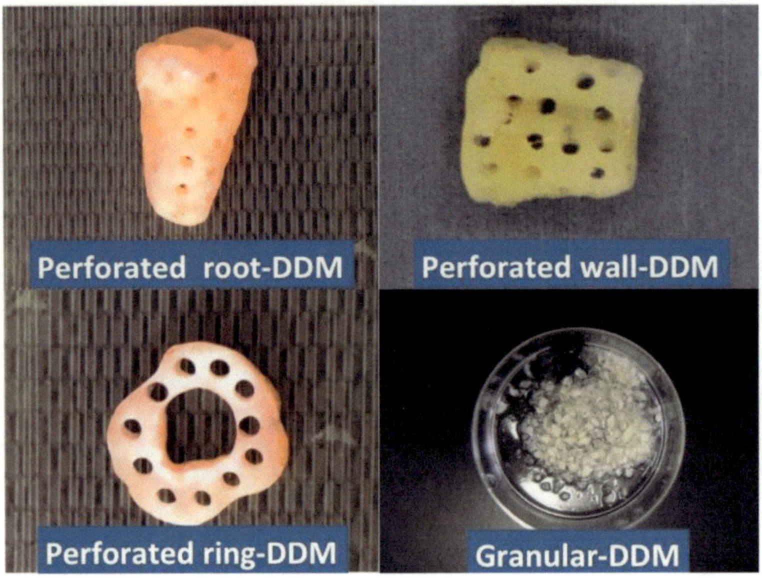

Figure 1.
Types of dentin materials.

However, cortical bone block inhibits the invasion of cells and blood vessels, because the calcified structure is highly dense like a castle wall [20].

3. History of dentin graft

First dentin autograft in human was achieved in 2002 in Japan for bone augmentation and reported in 2003, 81th IADR, Sweden [21]. The first clinical case was a sinus lifting, using demineralized dentin matrix (DDM) granules, derived from nonfunctional vital upper molars (#17, embedded #18) for implant placement [21]. We consider that the nonfunctional teeth are medical resources and can be recycled for transplant and bone regeneration. Recently, DDM graft has become a standard technique in South Korea. The unique service system in Korea Tooth Bank is the preparation and delivery of the tooth-derived materials on demand [4, 7, 8]. The tooth-derived materials were named as auto-tooth bone (ATB), which is divided into the block-type and powder-type [22]. The demineralized dentin, hydrated in 0.9% NaCl solution for 15–30 min before use, can be cut by operators with a surgical knife or scissors. We have thought of nonfunctional teeth as innovative resources and have advocated the medical recycle of the patient's own tooth as novel materials for bone regeneration. This matrix-based therapy is "*dental innovation,*"early in twenty-first century. Our advanced technique will expand from East Asia to the world, and dentin graft has been becoming a realistic alternative to the bone autograft.

4. Characteristics of dentin and bone

Dentin and bone are mineralized tissues. Dentin is cell-free matrix without a blood vessel, while bone includes osteocytes and vessels. However, dentin and bone are almost the same in chemical components. They consist of biological apatite (HAp: 70%), collagen (18%), non-collagenous proteins (NCPs: 2%), and body fluid (10%) in weight volume (**Table 1**). BMPs and FGFs are matrix-binding proteins,

	Inorganic	Organic	Fluid	BMP
Dentin	70	20	10	+
Cortical bone	70	20	10	+
Cartilage	2	28	70	+
Ceramics	100	0	0	–
Collagen	0	100	0	–

Table 1.
Chemical components (wt/v%) of human dentine and bone.

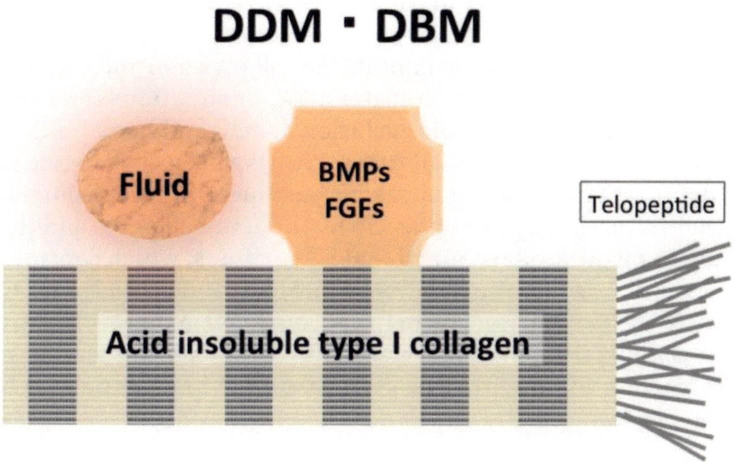

Figure 2.
DDM and DBM. BMPs and FGFs, matrix-binding proteins in NCPs. Collagen, mainly type I collagen.

while OCN is a mineral-binding protein in NCPs [13]. The small integrin-binding ligand, N-linked glycoprotein (SIBLING) family, including dentin sialophosphoprotein (DSPP), dentin matrix protein 1 (DMP1), bone sialoprotein (BSP), and osteopontin (OPN), are secreted into the ECM during the biomineralization [23–26]. The SIBLING phosphoproteins (DSPP, DMP1, BSP, OPN) bind to titanium beads [27]. DDM and DBM are mainly type I collagen (95%); the remaining is made up of NCPs, including a small amount of growth factors. Briefly, DDM and DBM are acid-insoluble collagen-binding bone morphogenetic proteins (BMPs) [28–30] and fibroblast growth factors (FGFs) [31–34] (**Figure 2**).

5. Advantages of demineralization

There are several advantages about demineralization process of dentin and bone. In our preparation, 2% HNO_3-demineralized dentin (DDM) is sterile material. The strong acid solution (pH < 1) can kill bacteria and decelluralize organ. Therefore, the strong acid treatment has antiseptic properties and decreases antigenicity. Bacteria-free DDM was detected after the culture for 7 days in blood agar medium of dentin granules demineralized in 2% HNO_3 for 20 min [35]. HAp crystals inhibit the release of BMPs along with growth factors [36]. After the removal of HAp crystals by ultrasonic demineralization, the surface area of dentin and bone increases remarkably [16]. In our study, human DDM and human DBM induced bone and cartilage independently at 4 weeks in the subcutaneous tissues of nude mice [34]. In addition, adult rat cortical bone plate treated with ultrasonic demineralization induced bone at 2 weeks, while fresh cortical bone plate never induces bone until 6 weeks [20]. These results indicated that highly calcified tissues such as cortical bone and calcified dentin did not have a better capability in bone induction than DDM and DBM. The delayed inductive properties of highly calcified dentin and bone may be related to the inhibition of BMP release by HAp crystals [36]. We never think, therefore, fresh bone is a gold standard. DDM and DBM have a better performance in bone induction than fresh dense bone.

6. Ultrasonic demineralization for dentin scaffold

Dense structure without pores inhibits the cell invasion and the body fluid permeation into the inside of the biomaterials. This situation is so-called material's wall. Material's wall means the exclusion of cells and body fluid. Dentin has compact structure with dentinal tubes. We have been challenging to create biological dentin scaffolds, using ultrasonic demineralization technology [14, 16, 37]. The whole structure design of dentin by the artificial pores and the ultrasonic treatment might produce functional 3D scaffolds, which control the bioabsorption rate and the adsorption ability for proteins and cells [2, 38] (**Figure 1**). The innovative technology can create the adequate 3D geometry and the surface structure of commercially available materials [14, 15] (**Figure 3**). Geometrical factors will improve the performance of biomaterials for bone regeneration [39–42].

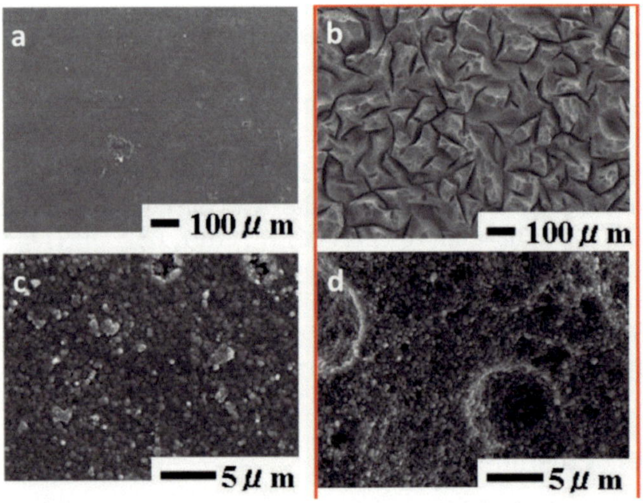

Figure 3.
SEM views of HAp dense plate before and after dissolution for 20 min in 2.0% HNO_3 by ultrasonic treatment at 600 W and 28 kHz. (a, c) Normal HAp dense plate (Cell-yard®); (b, d) ultrasonic demineralized plate; (a) dense flat plate; (b) knife cut-like grooves; (c) dense round crystals; (d) crater-like holes.

7. Biochemistry of DDM and DBM

Both DDM and DBM are composed of predominantly type I collagen (95%) and matrix-binding proteins such as BMPs [39, 43, 44]. BMPs bind to type I collagen of dentin and bone, even after complete demineralization (**Figures 1** and **2**). The fact is a reason why DDM and DBM induce bone and cartilage. Completely demineralized rabbit dentin matrix induced bone in the muscle at 4 weeks, while calcified dentin induced bone at 8–12 weeks after implantation [45, 46]. Many researchers made effort to discover dentin-derived BMPs [31, 47–49]. In 1990, BMPs, transforming growth factor beta (TGF-β), insulin growth factor-I (IGF-I), and IGF-II, were detected in human dentin [50]. Moreover, DDM and DBM possess the ability to coagulate blood plasmas [51]. The coagulation action of blood plasma by DDM should become advantageous for surgical operations. Interestingly, antibacterial activity within degradation products of biological scaffolds composed of extracellular matrix was published [52]. Additionally, extracellular matrix extracts from the dentin and pulp showed antibacterial activity against three types of anaerobic bacteria associated with dental disease [53].

8. Dentin scaffolds

DDM is defined as an acid-insoluble dentin collagen with natural cross-links and is a cell-free absorbable biomatrix with dentinal tube structure. DDM from autogenous tooth can be applied for bone grafts and tissue engineering as its own biomaterial, thus allowing improvement of bone induction while reducing the risk of infection. Human DDM can be recycled as cross-linked collagen in familial graft and allograft, because DDM is a cell-free matrix without antigenicity. We published bone regeneration in sheep iliac bone defect by human DDM root-type scaffold or β-TCP block [2, 54] (**Figure 4**). Bovine- and pig-derived collagenous materials are available as medical devices for the human body all over the world.

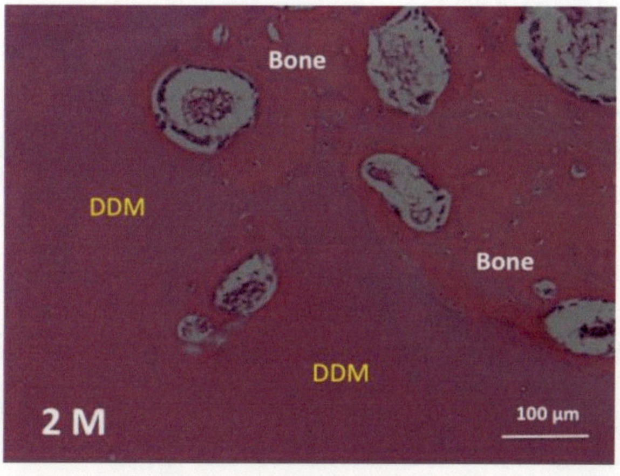

Figure 4.
Human DDM scaffold in sheep iliac bone defect at 2 months (HE).

9. Comparison of dentin scaffolds and biomaterials

Dentin scaffolds are composite materials of organic and inorganic components with natural cross-linking. Active 3D structure can be created in dentin scaffolds by using ultrasonic demineralization method and/or handmade technique under doctor's idea. Collagenous materials and ceramics are commercially available for medical use. Collagenous materials are a single component (mainly type I collagen) or a mix of type I collagen and gelatin. The collagen fibers are derived from bovine or pig. Hydroxyapatite (**Figure 3**), carbonate apatite, and β-TCP are bio-ceramics without growth factors and organic matter (**Table 1**).

10. Clinical case with DDM scaffolds for onlay graft on atrophic mandible

10.1 Patient

A 63-year-old female presented with missing teeth (#35-#37). Root apical legion was found in #34, and periodontitis was observed in #14 and 15 regions (**Figure 5a**). A clinical examination revealed highly atrophic bone in the mandible. Her medical history was unremarkable.

10.2 Surgical procedure 1

X-ray photos were taken before surgery. Non-vital teeth with lesion (#14, 15, 34) were extracted, and the debris and foreign body were removed carefully. Two teeth were prepared for root-type DDM (**Figure 5b**). The roots were perforated by a round bur, using the medical device (FIX®, Tokyo Iken Co., Ltd), and demineralized in 2% HNO_3 solution for 30 min. The other was crushed with saline ice by an electric mill

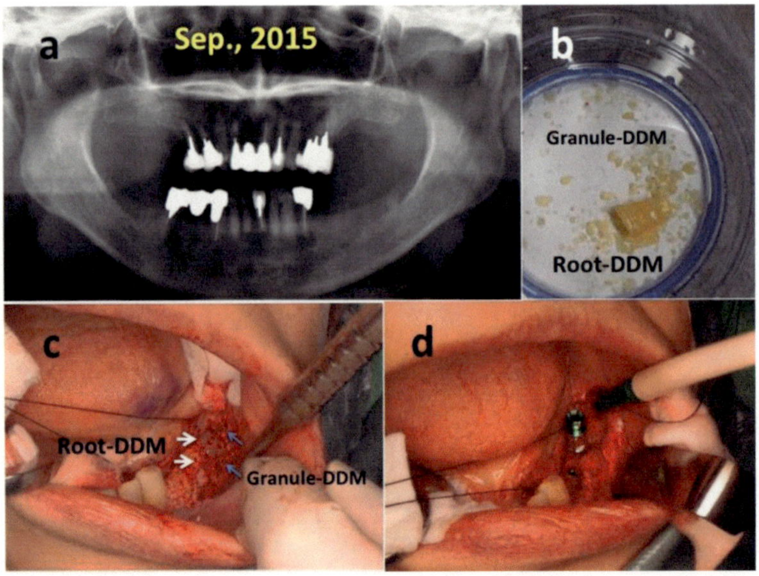

Figure 5.
Case: 63-year-old woman, lateral augmentation (#35 and 36 regions) by DDM. (a) Initial X-ray view; (b) DDM materials (roots, granules); (c) immediate DDM graft; note, lateral augmentation; (d) view after placement of fixtures.

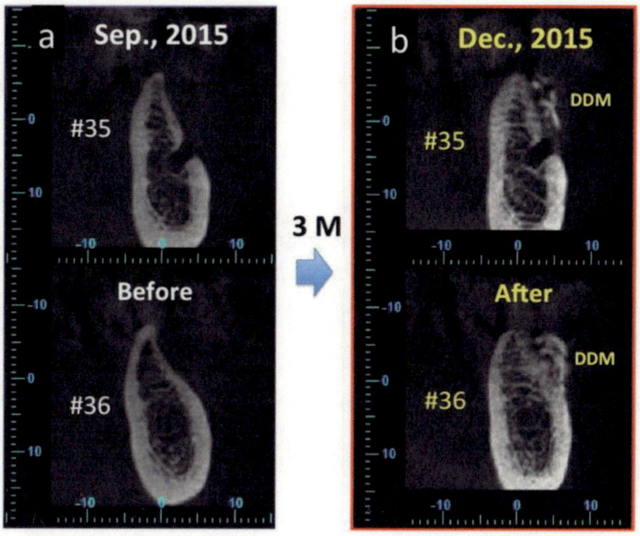

Figure 6.
X-ray CT before and after augmentation by DDM in 63-year-old woman. (a) Initial X-ray CT (#35, 36 region); note, ridge of atrophic bone; (b) 3 months after DDM graft; note, DDM combined with original bone.

(Osteo-Mill®, Tokyo Iken Co., Ltd) at 12,000 rpm for 30 s (Japan Patent: 4,953,276; USA Patent: 8,752,777). Next, the crushed tooth granules were decalcified in 2% HNO_3 for 20 min [55]. Two types of DDM were rinsed in cold distilled water. Perforations were performed into the atrophic bone (#35, 36) [43, 56]. DDM scaffolds were grafted for lateral augmentation (**Figure 5c**), covered by collagen membrane, and immediate implant placement (∅3.7 mm) was done into #34 socket under intravenous sedation.

10.3 Surgical procedure 2

X-ray CT was taken at 3 months after DDM graft. DDM residues did not get out during the drilling, and two fixtures (∅3.7 mm) were implanted into the augmented bone (#35, 36)(**Figure 5d**).

10.4 Results and discussion

The CT images at 3 months showed remarkable lateral augmentation, compared with that of preoperation (**Figure 6**). Non-vital teeth-derived DDM scaffolds were received to host, and two fixtures (∅3.7 mm) were placed into the DDM-augmented bone. Partially demineralized dentin matrix from non-vital tooth is acellular composite of collagen and hydroxyapatite without BMP activity. As original bone contains BMP activity, bone managements such as ultrasonic scaler treatment and perforations into cortical bone must be essential.

This patient was successfully restored with her own teeth and implant surgery. This case was a unique DDM onlay graft, combined with two types of DDM (roots, granules) for lateral bone augmentation in 2015.

11. Conclusion

Vital- or non-vital tooth-derived DDM can be recycled as natural scaffolds for local bone engineering. DDM graft is a matrix-based therapy without cells. Doctors can use patient's own teeth as bone graft materials. Biomaterial science should support and develop the advanced regenerative therapy using tooth-derived materials for patients in the near future.

Acknowledgements

This project was greatly supported by the grant-in-aid B (18H02997) of Japan Society for the Promotion of Science, the grant of Northern Advancement Center for Science & Technology (2017-2018), and the grant (consortium: 2004-5) of Japan Ministry of Economy, Trade and Industry.

Author details

Masaru Murata[1*], Naoto Okubo[2], Mamata Shakya[1], Md Arafat Kabir[1], Kenji Yokozeki[1], Bowen Zhu[1], Masahiro Ishikawa[3], Ryuji Kitamura[4] and Toshiyuki Akazawa[5]

1 Health Sciences University of Hokkaido, Japan

2 Hokkaido University, Japan

3 Ishikawa Dental Clinic, Japan

4 Toyohira Animal Hospital, Japan

5 Hokkaido Research Organization, Japan

*Address all correspondence to: murata@hoku-iryo-u.ac.jp

References

[1] Kabir MA, Murata M, Kusano K, Zakaria SM, Noor AHM, Khuda F, et al. Radiological evaluation of human dentin autografts in Bangladesh. Journal of Hard Tissue Biology. 2014;**23**:363-370

[2] Kabir MA, Murata M, Akazawa T, Kusano K, Yamada K, Ito M. Evaluation of perforated demineralized dentin scaffold on bone regeneration in critical-size sheep iliac defects. Clinical Oral Implants Research. 2017;**28**(11):e227-e235

[3] Kim YK, Kim SG, Byeon JH, Lee HJ, Um IU, Lim SC, et al. Development of a novel bone grafting material using autogenous teeth. Oral Surgery, Oral Medicine, Oral Pathology, Oral Radiology, and Endodontics. 2010;**109**(4):496-503

[4] Kim YK. Bone graft material using teeth. Journal of the Korean Association of Oral and Maxillofacial Surgens. 2012;**38**(3):134-138

[5] Kim YK, Lee JK, Kim KW, Um IW, Murata M. Healing mechanism and clinical application of autogenous tooth bone graft material. In: Pignatello R, editor. Advances in Biomaterials Science and Biomedical Applications. Croatia: InTecH Publisher; 2013. pp. 404-435. ISBN: 978-953-51-1051-4

[6] Kim YK, Lee J, Um IW, Kim KW, Murata M, Akazawa T, et al. Tooth-derived bone graft material. Journal of the Korean Association of Oral and Maxillofacial Surgeons. 2013;**39**(3):103-111

[7] Kim YK, Um IW, Murata M. Tooth bank system for bone regeneration—Safety report. Journal of Hard Tissue Biology. 2014;**23**:371-376

[8] Kim YK, Um IW, An HJ, Kim KW, Hong KS, Murata M. Effects of demineralized dentin matrix used as an rhBMP-2 carrier for bone regeneration. Journal of Hard Tissue Biology. 2014;**23**:415-422

[9] Kim YK, Bang KM, Murata M, Mitsugi M, Um IW. Retrospective clinical study of allogenic demineralized dentin matrix for alveolar bone repair. Journal of Hard Tissue Biology. 2017;**26**:95-102

[10] Kimura M, Murata M. Autotransplantation of teeth and decalcified dentin into atrophied anterior maxilla. In: Masaru M, Um WI, editors. Advances in Oral Tissue Engineering. Chicago: Quintessence Publishing Co. Inc; 2014. pp. 43-45

[11] Murata M, Akazawa T, Mitsugi M, Um IW, Kim KW, Kim YK. Human dentin as novel biomaterial for bone regeneration. In: Pignatello R, editor. Biomaterials—Physics and Chemistry. Croatia: InTech Publisher; 2011. pp. 127-140. ISBN: 978-953-307-418-4

[12] Murata M, Sato D, Hino J, Akazawa T, Tazaki J, Ito K, et al. Acid-insoluble human dentin as carrier material for recombinant human BMP-2. The Journal of Biomedical Materials Research. 2012;**100A**:571-577

[13] Murata M, Akazawa T, Mitsugi M, Kabir MA, Minamida Y, Um IW, et al. Autograft of dentin materials for bone regeneration. In: Pignatello R, editor. Advances in Biomaterials Science and Biomedical Applications. Croatia: InTech Publisher; 2013. pp. 391-403. ISBN:978-953-51-1051-4

[14] Akazawa T, Murata M, Tabata Y, Ito M. Bone Regeneration (ISBN) Chapter: Biomimetic Microstructure and Biocompatibility of Hydroxyapatite Porous Ceramics Designed by a Partial Dissolution-Precipitation Technique with Supersonic Treatment. Croatia: InTech Publisher; 2012. pp. 275-292

[15] Akazawa T, Murata M, Hino J, Nagano F, Shigyo T, Nomura T, et al. Surface structure and biocompatibility of demineralized dentin matrix granules soaked in a simulated body fluid. Applied Surface Science. 2012;**262**:51-55

[16] Akazawa T, Murata M, Minamida Y, Kabir A, Ito M, Katayama A, et al. Interface function design and bone-regenerative engineering of biomimetic biomaterials by supersonic treatment using electrolyzed water. Key Engineering Materials. 2015;**631**:241-246

[17] Akazawa T, Murata M, Minamida Y, Kabir A, Ito M, Sakamoto M, et al. Interface function and cefazolin-adosorption-release characteristics of hydroxyapatite granules modified by supersonic treatment techniques. Key Engineering Materials. 2016;**696**:265-270

[18] Solheim E. Osteoinduction by demineralised bone. International Orthopaedics. 1998;**22**:335-342

[19] Finkereier CG. Bone-grafting and bone-graft substitutes. The Journal of Bone and Joint Surgery. American Volume. 2002;**84**:454-464

[20] Shakya M, Yokozeki K, Akazawa T, Murata M. Rapid bone induction of cortical bone treated with ultrasonic demineralization in acidic electrolyzed water. Journal of Hard Tissue Biology. 2018;**27**:269-271

[21] Murata M. Autogenous demineralized dentin matrix for maxillary sinus augmentation in human. The First Clinical Report. 81th International Association for Dental Research, Geteburg, Sweden; 2003

[22] Park SM, Um IW, Kim YK, Kim KW. Clinical application of auto-tooth bone graft material. Journal of the Korean Association of Oral and Maxillofacial Surgens. 2012;**38**(1):2-8

[23] Fisher LW, Torchia DA, Fohr B, Young MF, Fedarko NS. Flexible structures of SIBLING proteins, bone sialoprotein, and osteopontin. Biochemical and Biophysical Research Communications. 2001;**280**:460-465

[24] Qin C, D'Souza R, Feng JQ. Dentin matrix protein 1 (DMP1): New and important roles for biomineralization and phosphate homeostasis. Journal of Dental Research. 2007;**86**:1134-1141

[25] Qin C, Brunn JC, Jones J, George A, Ramachandran A, Gorski JP, et al. A comparative study of sialic acid-rich proteins in rat bone and dentin. European Journal of Oral Sciences. 2011;**109**:133-141

[26] Sun Y, Lu Y, Chen S, Prasad M, Wang X, Zhu Q, et al. Key proteolytic cleavage site and full-length form of DSPP. Journal of Dental Research. 2010;**89**:498-503

[27] Kuboki Y, Furusawa T, Sato M, Sun Y, Unuma H, Abe S, et al. Bone enhancing effect of titanium-binding proteins isolated from bovine bone and implanted into rat calvaria with titanium scaffold. Bio-medical Materials and Engineering. 2014;**24**(3):1539-1548

[28] Huggins CB, Urist MR. Dentin matrix transformation: Rapid induction of alkaline phosphatase and cartilage. Science. 1970;**6,167**(3919):896-898

[29] Togari K, Mitazawa K, Yagihashi K, Tabuchi M, Maeda H, Kawai Y, et al. Bone regeneration by demineralized dentin matrix in skull defects of rats. Journal of Hard Tissue Biology. 2011;**21**(1):25-33

[30] Urist MR. Bone: Formation by autoinduction. Science. 1965;**150**:893-899

[31] Butler WT, Mikulski A, Urist MR, Bridges G, Uyeno S. Noncollagenous proteins of a rat dentin matrix

possessing bone morphogenetic activity. Journal of Dental Research. 1977;**56**:228-232

[32] Inoue T, Deporter DA, Melcher AH. Induction of chondrogenesis in muscle, skin, bone marrow, and periodontal ligament by demineralized dentin and bone matrix in vivo and in vitro. Journal of Dental Research. 1986;**65**:12-22

[33] Murata M, Kawai T, Kawakami T, Akazawa T, Tazaki J, Ito K, et al. Human acid-insoluble dentin with BMP-2 accelerates bone induction in subcutaneous and intramuscular tissues. Journal of the Ceramic Society of Japan. 2010;**118**(6):438-441

[34] Murata M, Akazawa T, Takahata M, Ito M, Tazaki J, Hino J, et al. Bone induction of human tooth and bone crushed by newly developed automatic mill. Journal of the Ceramic Society of Japan. 2010;**118**(6):434-437

[35] Fujita M, Uehara O, Miyakawa H, Naito S, Kabir A, Murata M, Nakazawa F. Detection of bacteria in human tooth-derived biomaterials. In: Masaru M, Um IW, editors. Advances in Oral Tissue Engineering. Chicago: Quintessence Publishing Co, Inc; 2014. pp. 31-32

[36] Huggins CB, Wiseman S, Reddi AH. Transformation of fibroblasts by allogeneic and xenogeneic transplants of demineralized tooth and bone. The Journal of Experimental Medicine. 1970;**132**:1250-1258

[37] Akazawa T, Murata M, Takahata M, Xianjun D, Abe Y, Nakamura K, et al. Characterization of microstructure and bio-absorption of the hydroxyapatite ceramics modified by a partial dissolution-precipitation technique using supersonic treatment. Journal of the Ceramic Society of Japan. 2010;**118**(6):535-540

[38] Kabir MA, Murata M, Akazawa T, Yamada K, Ito M, Shibata T. Demineralized dentin matrix with artificial perforations for bone regeneration in critical defects of adult sheep. Phosphorus Research Bulletin. 2016;**31**:1-3

[39] Akazawa T, Murata M, Sasaki T, Tazaki J, Kobayashi M, Kanno T, et al. Biodegradation and bioabsorption innovation of the functionally graded cattle-bone-originated apatite with blood compatibility. The Journal of Biomedical Materials Research. 2006;**76A**(1):44-51

[40] Kuboki Y, Saito T, Murata M, Takita H, Mizuno M, Inoue M, et al. Two distinctive BMP-carriers induce zonal chondrogenesis and membranous ossification, respectively; geometrical factors of matrices for cell-differentiation. Connective Tissue Research. 1995;**31**:1-8

[41] Murata M, Inoue M, Arisue M, Kuboki Y, Nagai N. Carrier-dependency of cellular differentiation induced by bone morphogenetic protein (BMP) in ectopic sites. International Journal of Oral and Maxillofacial Surgery. 1998;**27**:391-396

[42] Reddi AH. Bone matrix in the solid state: Geometric influence on differentiation of fibroblasts. Advances in Biological and Medical Physics. 1974;**15**:1-18

[43] Murata M, Maki F, Sato D, Shibata T, Arisue M. Bone augmentation by onlay implant using recombinant human BMP-2 and collagen on adult rat skull without periosteum. Clinical Oral Implants Research. 2000;**11**:289-295

[44] Murata M, Akazawa T, Tazaki J, Ito K, Sasaki T, Yamamoto M, et al. Blood permeability of a novel ceramic scaffold for bone morphogenetic protein-2. Journal of Biomedical Materials Research. 2007;**81B**(2):469-475

[45] Bang G, Urist MR. Bone induction in excavation chambers in matrix of

decalcified dentin. Archives of Surgery. 1967;**94**(6):781-789

[46] Yeomans JD, Urist MR. Bone induction by decalcified dentine implanted into oral, osseous and muscle tissues. Archives of Oral Biology. 1967;**12**:999-1008

[47] Bessho K, Tagawa T, Murata M. Purification of rabbit bone morphogenetic protein derived from bone, dentin, and wound tissue after tooth extraction. Journal of Oral and Maxillofacial Surgery. 1990;**48**:162-169

[48] Kawai T, Urist MD. Bovine tooth-derived bone morphogenetic protein. Journal of Dental Research. 1989;**68**:1069-1074

[49] Urist MR, Mizutani H, Conover MA, Lietze A, Finerman GA. Dentin, bone, and osteosarcoma tissue bone morphogenetic proteins. Progress in Clinical and Biological Research. 1982;**101**:61-81

[50] Finkelman RD, Mohan S, Jennings JC, Taylor AK, Jepsen S, Baylink DJ. Quantitation of growth factors IGF-I, SGF/IGF-II, and TGF-beta in human dentin. Journal of Bone and Mineral Research. 1990;**5**(7):717-723

[51] Huggins CB, Reddi AH. Coagulation of blood plasma of Guinea pig by the bone matrix. Proceedings of the National Academy of Sciences of the United States of America. 1973;**70**(3):929-933

[52] Brennan EP, Reing J, Chew D, Myers-Irvin JM, Young EJ, Badylak SF. Antibacterial activity within degradation products of biological scaffolds composed of extracellular matrix. Tissue Engineering. 2006;**12**(10):2949-2955

[53] Smith JG, Smith AJ, Shelton RM, Cooper PR. Antibacterial activity of dentine and pulp extracellular matrix extracts. International Endodontic Journal. 2012;**45**(8):749-755

[54] Murata M, Akazawa T, Kabir A, Minamida Y, Shakya M, Nagayasu H, et al. Highly porous β-TCP block with triple pore structure in rat subcutaneous tissue and sheep iliac critical bone defect. Key Engineering Materials. 2016;**696**:187-191

[55] Murata M, Akazawa T, Tazaki J, Ito K, Hino J, Kamiura Y, et al. Human dentin autograft for bone regeneration–automatic pulverizing machine and biopsy. Bioceramics. 2009;**22**(22):745-748

[56] Murata M, Huang BZ, Shibata T, Imai S, Nagai N, Arisue M. Bone augmentation by recombinant human BMP-2 and collagen on adult rat parietal bone. International Journal of Oral and Maxillofacial Surgery. 1999;**28**:232-237

Chapter 3

Advanced Surface Treatments on Titanium and Titanium Alloys Focused on Electrochemical and Physical Technologies for Biomedical Applications

Elena Fuentes, Sofia Alves, Ainara López-Ortega, Lucía Mendizabal and Virginia Sáenz de Viteri

Abstract

Titanium and its alloys are becoming very promising materials in biomedicine due to their excellent properties. However, their poor tribological behavior characterized by high friction coefficient and severe adhesive wear is their main limitation. Surface modification technologies based on electrochemical and physical techniques have been successfully employed to improve the tribological performance and osseointegration of Titanium materials, ensuring an effective protection against both wear and corrosion. For instance, anodizing and plasma electrolytic oxidation (PEO) are two electrochemical techniques that allow the growth of an oxide film of high hardness and good adhesion. The formation of these oxide films in electrolytes with bioactive elements has been reported to enhance cell functionalities and improve the tribocorrosion performance of Titanium surfaces considerably. Similarly, physical vapor deposition (PVD) technologies such as cathodic arc evaporation (CAE) and magnetron sputtering (MS) are commonly used today for the growth of protective hard coatings on different Titanium components in the biomedical field. Diamond-like-carbon (DLC) and transition metal nitride (MeNx) and carbide (MeCx) protective films grown by PVD have proven to be excellent candidates to enhance Titanium and Titanium alloys performance and durability, owing to their excellent adhesion, high hardness, low friction coefficient and enhanced wear and corrosion resistance.

Keywords: Titanium, Titanium alloys, surface treatments, anodizing, plasma electrolytic oxidation, physical vapor deposition, osseointegration

1. Introduction

Biomaterials are of great importance in medicine because of their ability to treat, improve, or replace damaged tissues, organs of body functions. Apart from ceramics, polymers, and composites, metallic materials rank distinguished in the field of biomaterials. Titanium (Ti) is considered the most biocompatible of all metals because of its superior resistance to corrosion from bodily fluids, bio-inertness, capacity for osseointegration, and high fatigue limit. Titanium develops a very strong passivating oxide layer which forms naturally in the presence of oxygen on its surface. This protective oxide film, with a thickness of 3–10 nm [1] is strongly adhered, insoluble, and chemically impermeable, preventing reactions between the metal and the surrounding environment.

Titanium was first used in surgery in the 1950s and in dentistry a decade earlier. Recently, Titanium-based materials are attracting much interest as implantable materials because of their superior corrosion resistance, better mechanical properties such as remarkably high specific strength, low elastic modulus, and excellent biocompatibility compared to other competing biomaterials like stainless steel, Co-Cr alloys and nitinol alloys. They are now extensively and routinely accepted by medical professionals as the material of choice for prosthetics, internal fixation, inner body devices and instrumentation.

Ti6Al4V is the most commonly used Titanium alloy in orthopedic applications, for knee and hip prosthesis bone screws, and plates. It is as strong as steel and twice as strong as aluminum, but it is 45% lighter than steel and only 60% heavier than aluminum. This alloy is widely used as hard-tissue replacements in artificial bones and joints because of their outstanding characteristics such as high strength, low density, immunity to corrosion, complete inertness to body environment, enhanced compatibility, relatively low Young's Modulus and high capacity to join with bone or other tissues. However, the most important limitation of this and other Titanium alloys is their poor tribological behavior characterized by a high friction coefficient, and a low wear resistance, suffering from a severe adhesive wear [2].

Research and development on Titanium's medical applications are concentrated on new alloys, production technologies and surface treatments that improve biocompatibility and prevent fretting fatigue. Nowadays, the relatively poor tribological properties and possible corrosion problems of orthopedic devices made of Titanium alloys have led to the development of suitable surface treatments to effectively increase near-surface strength, improving the hardness and abrasive/adhesive wear resistance, thereby reducing the friction coefficient as well as avoiding or reducing the transference of ions from the surface or bulk material to the surrounding tissue, and improving biocide capacity and osseointegration of implants. Some surface treatment methods prone to be applied to achieve these objectives could be mechanical, chemical, electrochemical and physical methods. Mechanical methods such as machining, grinding, polishing and blasting produce specific surface topographies to improve adhesion in bonding. Chemical methods include acidic or alkaline treatments, hydrogen peroxide treatment and sol-gel, whose main objective is to improve biocompatibility, bioactivity or bone conductivity as well as chemical vapor deposition (CVD) for improving wear and corrosion resistance and blood compatibility, and biochemical techniques for inducing specific cell and tissue responses. Electrochemical methods include anodic oxidation, cathodic deposition and plasma electrolytic oxidation for improving corrosion resistance, biocompatibility, bioactivity or osseointegration. Physical methods involve thermal spray, physical vapor deposition (PVD), ion implantation/deposition and glow discharge plasma treatments for improving mainly wear and corrosion resistance.

In this chapter, two very promising surface modification methods, electrochemical and physical techniques, have been considered to develop highly corrosion- and wear-resistant and totally biocompatible coatings with improved osseointegration properties that extend the performance of Titanium alloys-based systems, several times beyond its natural capacity.

Electrochemical techniques such as anodizing and plasma electrolytic oxidation (PEO) have been employed to synthesize well-controlled ceramic-like oxide

TiO_2 films with improved features. Both techniques are simple, cheap and effective electrolytic passivation processes used to increase the thickness of the natural oxide layer on the surface of Titanium-based materials parts. In both techniques, a biocompatible and corrosion resistant oxide layer of high hardness and good adhesion is generated on the surface of the alloy. In anodizing, the part to be treated forms the anode electrode of an electrolytic cell. A direct current is passed through the part to be anodized while submerged in a water-based electrolyte. The water breaks down, liberating oxygen at the surface of the part, which then combines with the Titanium to form the thick Titanium oxide layer. Thickness is determined by the level of electrical current and the process duration applied. In PEO method, the layers are formed by polarizing the Titanium-based part to the dielectric breakdown voltage in a suitable electrolyte. A wide range of polarization conditions are available for formation of the coatings, including DC (direct current) and AC (alternating current), with control of the current, voltage or power supplied to the cell. The main differences between anodizing and PEO techniques rely on the lower voltage that is used for anodizing with no discharges/plasma generation during the process, and higher thickness in case of PEO. Despite the similarities between both, the lower voltage applied during anodizing significantly influences the properties of the anodic film grown during the process in what regards morphology, topography, chemistry and crystalline structure. The formation of Titanium oxides by anodizing in electrolytes composed of bioactive elements, has been reported as a very promising method either to enhance cell functionalities or to improve the tribocorrosion performance of Titanium surfaces. The ceramic oxide layers generated by PEO technique are characterized by a high corrosion and wear resistance. The oxide layers generated by PEO on Titanium alloys have been observed to have a suitable topography for cellular proliferation, and improve the wear-corrosion response of the substrate considerably.

As well, physical vapor deposition (PVD) technologies are commonly used today for the growth of protective hard coatings on different Titanium components in the biomedical field. Among them, cathodic arc evaporation (CAE) and magnetron sputtering (MS) techniques are the most popular ones. Diamond-like-carbon (DLC) and transition metal nitride (MeN_x) and carbide (MeC_X) protective films grown by PVD techniques have proven to be excellent candidates to enhance Titanium and Titanium alloys performance and durability in biomedical applications, owing to their excellent adhesion, high hardness, low friction coefficient and enhanced wear and corrosion resistance in body fluids. Besides this, PVD process parameters can be tuned to develop biomedical protective coatings with tailored properties. Particularly, Ti-C-N, Ti-C-N + Ag, Ti_DLC, (Ti,Zr)CN, TaN and TaN_Ag films performance to enhance Titanium and Titanium alloys features is described in this chapter, but the number of possible PVD coating materials (including completely different characteristics) is huge and continuously increases.

2. Problematics related to Titanium alloys in biomedical implants

A successful clinical implantation depends on the initial primary stability provided by the distribution of bone tissue around the implant, its quality and its amount. The integration of the implant in the bone is affected by several factors [1, 3–9]. The implant material, the design characteristics of the implant and its surface features, the implant loading conditions, the surgical technique employed, or the microbial adhesion and colonization are some examples. Also, the state of the host, i.e., the quantity and quality of the bone, and the mismatch of the mechanical properties of the bone and the implant are of great importance. Finally, the release of the wear particles of metal ions from the implant during implantation is a key issue concerning the durability of the implants. In this context, wear, corrosion, and their synergistic interaction is a concerning issue to deal with, once it might influence the durability and clinical success of biomedical implants, and this chapter provides a brief insight into these phenomena and their detrimental effects.

2.1 Wear

One common degradation mechanism of orthopedical materials is wear. The integration of the implant in the bone tissue involves a relative movement between both surfaces. Furthermore, once the implant has been placed, the loads generated by the daily activity of the human body will result on micromovements on the implant/bone interface (fretting) [7, 10, 11]. One of the principal drawbacks of Titanium alloys relay on their low fretting-fatigue resistance and their poor tribological properties because of their low hardness [12–14] and the poor mechanical integrity of the TiO_2 passive film formed on their surface [15, 16]. Titanium-based alloys are characterized by a high coefficient of friction and severe wear either against themselves or other materials [2, 17]. Titanium has tendency for moving or sliding parts to gall and eventually seize. This causes a more intensive wear as a result of creation of adhesion couplings and mechanical instability of passive layer of oxides, particularly in presence of third bodies. This implies the liberation of metallic particles or wear debris from the implant material into the surrounding tissues that results in an inflammation and gives rise to the bone resorption (osteolysis), which ultimately leads to loosing of the implant and hence the implant has to be replaced by a new one.

2.2 Corrosion

On the other hand, the human body comprises a highly corrosive environment, which might accelerate the degradation of metallic implants, regardless of their high corrosion resistance [18]. Corrosion is a degradation mechanism resulting from the electrochemical processes of oxidation and reduction taking place on a metal in a hostile electrolytic environment [19, 20]. The corrosion characteristics of implant alloys are influenced by the passive film that is formed on their surface [8]. This passive layer, of about 3–10 nm [1] and composed of TiO_2, is the responsible for the adequate corrosion resistance of Titanium alloys, by acting as a physical barrier between the metal and the surrounding environment [21]. However, the presence of chloride ions (Cl^-) makes human fluids considerably aggressive. Due to the small size of these ions, they can penetrate through the passive film and break down this protective layer by inducing localized corrosion [9, 22]. The corrosion of Titanium alloys has been found to be influenced by several factors, either accelerating or inhibiting their degradation. Some examples are the pH levels [23], the presence of proteins [24], the presence of bacterial colonies [25], or human cells [26], among others. Corrosion of biomedical implants would lead to the liberation and accumulation of metallic ions and corrosion products [10, 17, 27].

2.3 Tribocorrosion

Since implants are simultaneously subjected to wear and corrosion solicitations, the degradation mechanism expected is a combination of both phenomena [28, 29]. When wear and corrosion take place simultaneously, the process is known as tribocorrosion [20, 30, 31]. Tribocorrosion involves a complex synergism between wear and corrosion, since the material loss when both processes take place together is higher than when they occur alone [10, 32–36]. This synergism is schematically represented in **Figure 1**. When the passive layer generated on the surface of the alloy is destroyed and removed (depassivation), fresh material is exposed to the electrolyte where corrosion reaction take place (wear-accelerated corrosion). On the other hand, the wear debris generated can oxidize and accumulate in the tribological contact increasing the wear extent (corrosion enhanced wear). The material degradation by tribocorrosion depends on different variables, such as the properties of the materials involved in the systems (composition, microstructure, roughness, mechanical properties...), the electrochemical parameters (composition of the corrosive media, pH, conductivity, temperature...), and the mechanical parameters (load, frequency of the contact, vibrations...).

All in all, the release of metallic ions and corrosion products coming from both wear and corrosion and their interaction will enhance the likelihood of biological complications. This can result in peri-implant inflammatory reactions and aseptic osteolysis leading to the loss of the implant [16, 37, 38]. Therefore, tribocorrosion is one significant clinical issue that can compromise the osseointegration process as well as the further mechanical integrity and the biomechanical stability of the implant [21, 39, 40].

2.4 Infection

Certain microorganisms can grow in layers, forming biofilms on medical surfaces like implants. Biofilm-associated infections on implants are responsible for 15–25% of implant failures as biofilms are resistant to most of the conventional anti-microbial agents. So far, implant coatings with biocidal properties have been generated, which release silver ions or conventional antibiotics to inhibit biofilm formation. The need to resolve biofilm-associated infections is urgent to reduce the necessity for revision surgery. Besides, higher medical costs such infections lead to significant pain and distress in patients.

2.5 Challenges of Titanium in biomedicine

For all the aforementioned reasons, a primary goal on implantology research is the mitigation of wear and corrosion debris formation during the implantation. For this aim, the improvement of the wear-corrosion resistance of the implant material is a key factor. Functionalization of implant surfaces has been recognized as a promising approach to overcome current problematics [41]. In this context, the use of surface modification techniques that allow the modification of features such as

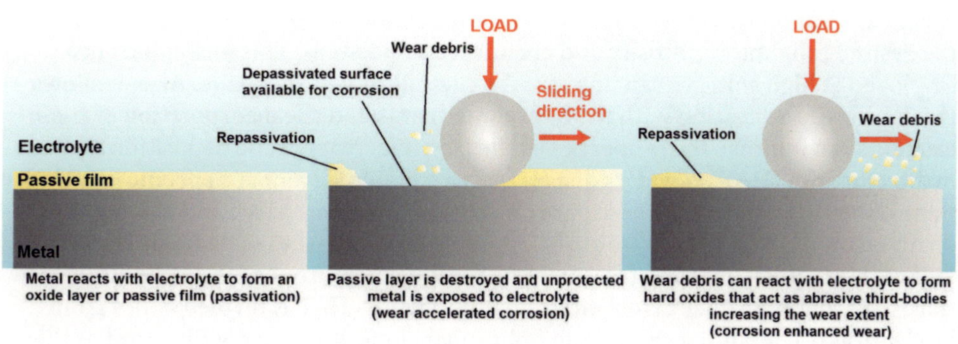

Figure 1.
Schematic representation of the synergism in tribocorrosion.

morphology, topography, structure, or chemistry has been gaining attention of the biomedical sector. In this sense, electrochemical techniques are simple, cheap and effective alternative methods used to improve the tribological and electrochemical properties of Titanium-based materials for biomedical applications, increasing their durability.

3. Electrochemical surface modification techniques

3.1 Introduction

Degradation of Titanium-based implant materials is feasible to occur *in vivo* conditions through corrosion and tribocorrosion processes, therefore being an issue of high clinical significance. One must be expected that the release of wear particles/ corrosion products through tribocorrosion processes taking place during implantation, may seriously compromise the osseointegration process [41]. Degradation and inflammatory processes may also contribute to the loss of mechanical integrity, influencing the magnitude of micromovements at implant/bone interface, thus compromising the biomechanical stability of the implant as well as the long-term health of peri-implant tissues [39, 40, 42]. Wear debris resulting from tribocorrosion, besides trigger locally aggressive biological reactions at peri-implant tissues, can go into bloodstream and subsequently be disseminated into human organs, leading to adverse effects at a systemic level, and therefore presenting serious risks for human health [43]. To overcome these problems, surface treatments have been widely employed to Titanium and its alloys, among which anodic oxidation (anodizing) at low and high voltage processing conditions is found.

The high biocompatibility and corrosion resistance of Titanium and its alloys are associated with the ability of Titanium to spontaneously form a stable and adherent TiO_2 thin film on its surface, named as passive film, when exposed to oxidizing conditions [42]. The characteristics of this oxide film of around 3–10 nm thick [1], such as its morphology, topography and chemical composition play an important role on its osseointegration ability [44, 45]. Titanium oxide, i.e., Titania (TiO_2) can be found on three polymorphic forms: rutile, anatase, and brookite. Anatase and brookite can convert to rutile upon heating, achieving the unique properties of rutile at room temperature. Recently developed high voltage techniques allow the generation of oxide films composed of crystalline rutile/anatase phases [8], which play a crucial role on their tribocorrosion responses [46].

Among the many methods of Titanium surface modification, electrochemical techniques are simple, cheap and effective. While anodizing is an anodic electrochemical technique, electrophoretic and cathodic depositions are cathodic electrochemical techniques. By anodic oxidation it is possible to obtain oxide films with the desired roughness, porosity and chemical composition. This technique once employed at high voltages can improve the crystallinity of the oxide. Also, it allows the doping of the coating with the bath constituents and the incorporation of these elements can improve the properties of the film. Electrophoretic deposition may use hydroxyapatite (HAP) powders dispersed in a suitable solvent at a specific pH. Under these operating conditions these particles acquire positive charge and coatings are obtained on the cathodic Titanium by applying an external electric field. These coatings require a post-sintering treatment to improve their properties. Cathodic deposition is another type of electrochemical method where HAP is formed *in situ* from an electrolyte containing calcium and phosphate ions. It is also possible to alter the structure and/or chemistry of the obtained deposit. Nano-grained HAP has higher surface energy and greater biological activity and therefore emphasis is being laid to produce these coatings by cathodic deposition [47].

As abovementioned, electrochemical techniques such as anodizing and plasma electrolytic oxidation (PEO) have been employed to synthesize ceramic-like oxide films with improved features. These treatments have been claimed to enhance the tribocorrosion behavior of Titanium-based materials, as well as their osseointegration by the incorporation of certain chemical elements into the oxide composition (e.g., calcium, phosphorus, zinc, etc.). In this section, anodizing and PEO processes are described, and some interesting findings on their tribocorrosion and biological performances are reported.

3.2 Anodizing

3.2.1 Definition and influence of anodizing parameters on anodic oxide film features

Anodizing has attracted considerable attention as a simple and fast electrochemical method to modify Titanium surface features, by promoting the growth of an oxide film on its surface [37, 48, 49]. Conventional anodizing process relies on the application of a predetermined current or voltage between Titanium material (anode) and an unreactive/inert material (cathode, e.g., platinum and graphite), previously immersed in a conductive solution. When polarized, a flux of electrons is generated between both materials, and the oxide film is grown on Titanium surface through oxidation reactions, along with field-driven ion diffusion [50–52].

The anodizing process is rather complex and depends on various parameters such as applied current density, concentration and composition of the electrolyte, the electrolyte temperature, duration of the process, the agitation speed of the solution during treatment and the cathode to anode surface area ratios [53]. Investigations have been carried out demonstrating that the features of Titanium oxide films generated by anodizing (e.g., morphology, topography, chemical composition and thickness) are strongly dependent on processing conditions, as below described in this section. In particular, over the last decades, many efforts have been done to develop nanoscale features on Titanium substrates to mimic the micro/nanostructures of natural bone and so, increase implants biocompatibility. In this field, anodizing technique is receiving a growing importance because of its convenience to create surfaces with biologically-inspired topographies with different chemical compositions according to the electrolyte used during the process [54–56].

Among various nanostructured materials, the decoration of Titanium-based surfaces with well-aligned TiO_2 nanotubes (NTs) has received special attention. Among the existing methods for TiO_2 NTs synthesis, anodizing has emerged as one of the most effective due to its versatility, easy operation, and commercial feasibility [52, 57, 58]. This technique is already widely used in industry to produce large-scale low-cost protective oxide coatings [52]. The anodizing process for NTs synthesis is commonly carried out by applying a constant voltage of 1–30 V in aqueous electrolytes, or 5–150 V in non-aqueous electrolytes, containing 0.1–1 wt.% fluoride ions (F^-) [51]. The anodizing setup for synthesis of TiO_2 NTs is illustrated in **Figure 2(a)**. The mechanisms of TiO_2 NTs growth by anodizing in a fluoride containing electrolyte, have been conventionally accepted as a result of a field assisted dissolution process consisting of three main stages: (1) the oxidation of Titanium metal involving Ti^{4+} ions formation/liberation; (2) the growth of an oxide film on Titanium surface, through recombination of Ti^{4+} and O^{2-} ions (provided by deprotonation of H_2O or OH^-) moving under the action of the electric field; (3) the local

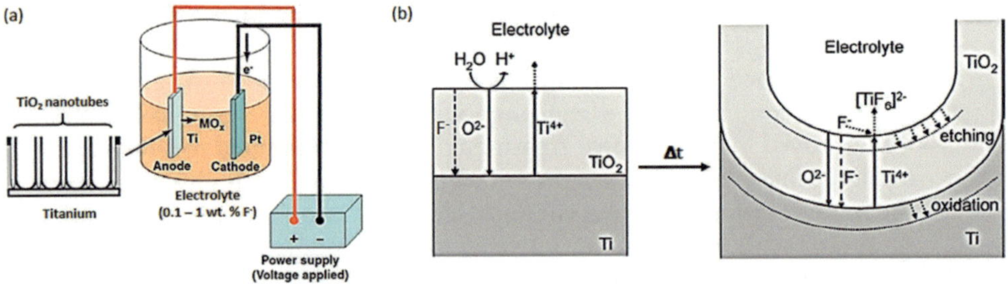

Figure 2.
(a) Anodizing setup for synthesis of TiO_2 NTs in a fluoride containing electrolyte. In (b) the mechanisms underlying nanotube growth evolution by anodizing are shown. Adapted from Refs. [51, 60, 61].

chemical dissolution of the growing oxide by fluoride ions and subsequently pore nucleation, with formation of water-soluble $[TiF_6]^{2-}$ species. This process assumes that nanotubes growth takes place through the balance established between the formation of the oxide film and its enhanced dissolution at the base of the pores/tubes, where the electric field is stronger, as schematically depicted in **Figure 2(b)** [50–52, 57, 59, 60].

The resulting features of TiO_2 NTs, namely morphology, chemistry, and length, may vary over a wide range according to the processing anodizing parameters, such as the applied voltage, duration, and electrolyte composition. As reviewed by Roy et al. [51], anodizing time and etching rate define tube length, while nanotube diameter is controlled linearly by the applied voltage. In general, thinner films are produced in acidic electrolytes as compared to neutral, mostly ascribed to the faster dissolution rate of the oxide in a lower pH solution [57]. Accordingly, Macak et al. [62] reported that by using a neutral NaF-based electrolyte, significantly thicker porous layers were obtained than in acidic solution. Furthermore, Shankar et al. [63] showed that by using non-aqueous organic electrolytes containing fluoride ions, such as ethylene glycol, highly ordered TiO_2 nanotube arrays up to 220 μm in length were grown, and depending on the anodic voltage, the inner pore diameters ranged from 20 to 150 nm. A remarkable advance has been reported by Han et al. [64], who fabricated TiO_2 nanotube arrays with enhanced self-ordering level by multistep anodic oxidation of Titanium in an organic electrolyte containing fluoride ions, and this methodology has been widely adopted for self-templating anodizing processes [65–67]. Generally, the structure of TiO_2 NTs grown by conventional anodizing is amorphous, however, it can become crystalline (e.g., anatase or a mixture of anatase and rutile) by thermal treatments at 280–800°C, for 2–3 h. Although this method is currently used to achieve enhanced surface bio-functionality, it is not standardized yet [51, 68].

3.2.2 Modification of Titanium and Titanium-alloys surface features by anodizing to improve the tribocorrosive and biological performances of osseointegrated implants

To achieve a good and fast osseointegration both the surface chemistry and topography of the implant material are of upmost importance. As abovementioned, various attempts have been made for the development of surfaces mimicking the hierarchical structure of bone which varies from micro- to nano-scale structures together with the inclusion of bioactive elements [69]. Zhao et al. [70, 71] observed that the conjunction of micro and nano-topographies on Titanium surfaces had a synergistic role on multiple cell functions, through the enhancement of multiple osteoblast functionalities, as compared to micro-textured surfaces. Furthermore, micro/nano-textured surfaces showed enhanced ability to induce mesenchymal stem cells (MSCs) osteogenic differentiation. The benefit of a hierarchical micro/nano-topography on promoting MSCs adhesion was also demonstrated by Zhang et al. [72].

In particular, anodic nanotubular structures made out of TiO_2 have demonstrated unique morphological and physicochemical features with potential to stimulate implant-bone integration, either *in vitro* or *in vivo,* by modulating osteoblasts and human MSCs functions such as adhesion, proliferation, and differentiation [73–75], when compared to conventional Titanium surfaces. Furthermore, TiO_2 NTs have demonstrated ability, not only to control bone-forming cell functions, but also other types of cells that might have a major role in osseointegration process such as, osteoclasts, endothelial cells, and immune system cells (e.g., macrophages) [76, 77]. An additional exciting feature of TiO_2 NTs is their ability to prevent microbial adhesion and colonization, and therefore to avoid implant-related infections, which may end up in failure [78–80]. As demonstrated by Ercan et al. [80], Titanium surface features including chemistry, crystallinity, nanotube size, and hydrophilicity, significantly influence the responses of *S. epidermidis* and *S. aureus* pathogens. Several studies have reported that nanotube diameter besides influence the microbial adhesion and colonization, also modulate biological responses [81–83]. Furthermore, TiO_2 NTs display an improved corrosion resistance as compared to Titanium-based surfaces [84, 85]. Demetrescu et al. [55] studied the electrochemical behavior of TiO_2 NTs in Fusayama's artificial saliva (AS) and concluded that very low corrosion current densities were recorded for TiO_2 NTs due to a strong passive oxide film formation. Electrochemical impedance spectroscopy (EIS) results indicated that TiO_2 NTs consisted of a bi-layered oxide made up of an inner barrier layer associated to high impedance and responsible for corrosion protection, and a porous outer layer (NTs) of lower impedance.

An additional particularity of TiO_2 NTs is their potential to behave as a platform for drug-eluting and local delivery, due to their excellent controllable dimensions, surfaces chemistry and large surface-to-volume ratio. Easily by changing the nanotube diameter, wall thickness, and length, the kinetics of specific drugs can be adjustable to achieve stable and controlled release [86–88]. Bio-functionalization of TiO_2 NTs may be easily achieved by conventional techniques already used for surface modification of biomaterials. In this way, TiO_2 NTs may behave as effective bio-selective surfaces by incorporating different elements in its structure, hence inhibiting bacterial functions and concomitantly promoting osteoblast responses. Recently, Alves et al. [89] have also synthesized anodic TiO_2 NTs incorporating Ca- and P-based compounds, through a novel methodology that resembles on reverse polarization anodizing processes in an electrolyte composed of those bioactive species. Results showed that hydrophilic Ca/P-NTs improved osteoblast-like cells adhesion and proliferation, as compared to Titanium smooth surfaces. Furthermore, the corrosion resistance of these surfaces in AS was significantly improved, when compared to conventional NTs and Titanium surfaces. A step forward, aiming to address, simultaneously, the lack of tissue integration and infection problems, Alves et al. [90] incorporated Zn in TiO_2 NTs by anodizing of nanotubular structures in an electrolyte containing Ca, P and Zn elements. Results show that bio-functionalized TiO_2 nanotubular surfaces are biocompatible and modulated cell morphology. In particular, NTs enriched with Ca, P, and Zn, induced to significantly up-regulated levels of bone morphogenetic protein 2 (BMP-2) and osteopontin (OPN) genes of human MSCs, when compared to conventional NTs. TiO_2 nanotubular surfaces induced human MSCs to release a higher amount of vascular endothelial growth factor (VEGF), and showed the ability to impair *S. aureus* viability, therefore behaving as bio-selective surfaces.

The growth of TiO_2 NTs on Titanium surfaces has demonstrated potential to reduce stress shielding effect, in accordance with their mechanical properties determined by nanoindentation. An elastic modulus of 4 – 43 GPa has been determined for TiO_2 nanotubular structures [91–93], which is much closer to that of natural bone (i.e., 0.02 – 30 GPa) [94–96] when compared to Titanium (i.e., 143 ± 23 GPa) [27, 97]. The mechanical behavior of TiO_2 NTs was studied by Xu et al. [93] by nanoindentation. In summary, the authors observed that TiO_2 NTs break as long as the indentation depth increases, interacting with neighboring NTs and causing them to bend and fracture, with formation of smaller fragments that become gradually compacted, resulting in densification. Furthermore, Crawford et al. [91], observed that TiO_2 NTs inelastically deform by tube crushing in the immediate vicinity of the indenter tip, accompanied by local densification. However, these studies provide very limited information to predict the *in vivo* degradation mechanisms that TiO_2 NTs undergo, when submitted to tribocorrosive actions. To emphasize the importance of this knowledge, it has been reported that nanotubular films are prone to peeling off from the Titanium substrate due to poor adhesion between them [98–100]. Therefore, tribocorrosion studies become of fundamental importance, not only because of the high probability of adhesion failure of the nanotubular systems, but also because of their high clinical relevance, since implant degradation *in vivo* can seriously compromise osseointegration or even its long-term stability. In the attempt to construct effective nanotubular systems based on an integrated and multidisciplinary approach that addresses the development of tribocorrosion resistant implant surfaces with potential to avoid infection, and simultaneously, promote osseointegration, Alves and co-workers have recently reported, for the first time, important findings on the tribocorrosion field [101, 102].

Alves et al. [101] studied the tribo-electrochemical behavior of bio-functionalized TiO_2 NTs in AS, providing a first insight on their degradation mechanisms under reciprocating sliding conditions. Titanium surfaces decorated with TiO_2 NTs of 50–90 nm diameter were bio-functionalized by anodizing in an electrolyte containing Ca, P and Zn. The results showed that the tribo-electrochemical behavior of TiO_2 NTs was significantly improved after bio-functionalization treatments, which was correlated with their improved adhesion strength to the Titanium substrate, granted by the formation of a nano-thick oxide film at the interface region. The authors found out that after conventional anodizing process used for NTs growth, there was the formation of a non-continuous interface characterized by a hollow space between the nanotubular film and the Titanium substrate. After bio-functionalization treatments by reverse polarization anodizing, remarkable changes were observed at the Ti/TiO_2 NTs interface region, due to the formation of a nano-thick oxide film (230–250 nm) during the second step anodizing process, which appeared to improve NTs adhesion to Titanium [103], as depicted in **Figure 3**. The adhesion properties and degradation mechanisms of TiO_2 NTs were investigated by tribo-electrochemical tests carried out in AS under single and multiple sliding actions, to better mimic real *in vivo* conditions that dental implants might be exposed to in real life [101, 102]. From these studies the poor adhesion of conventional TiO_2 NTs was undoubtedly confirmed, and bio-functionalization came out as a very promising approach to overcome it. Besides the more active electrochemical state, conventional TiO_2 NTs suffered catastrophic destruction right after 100–300 s sliding actions accompanied by significantly higher wear volume loss as compared to bio-functionalized NTs, either when submitted to single or multiple sliding actions in AS. The improved tribo-electrochemical behavior after bio-functionalization was correlated with the significantly higher adhesion strength of the NTs to the substrate, granted by the nano-thick interfacial film formed by anodizing.

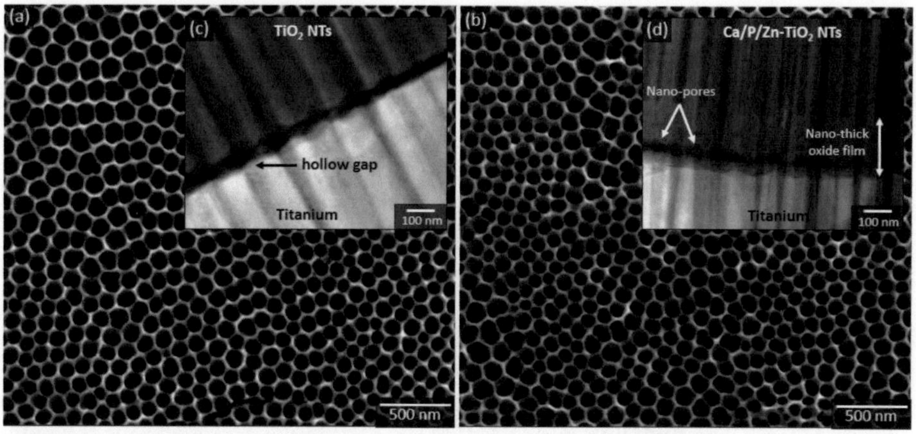

Figure 3.
SEM micrographs of (a) NTs and (b) NT-Ca/P/Zn surfaces. Dark-field STEM micrographs at the interface of (c) NTs and (d) NT-Ca/P/Zn nanotubular films. In (c) the inset arrow shows the hollow gap between Titanium substrate and NT film while in (d) shows the nano porosity at the interface as well as the nano-thick oxide film grown during bio-functionalization (230–250 nm) [102].

The high adhesion strength was correlated with the higher hardness measured for these films, which consequently enhanced their mechanical wear resistance. From these studies a first insight on the main degradation mechanisms of TiO_2 NTs was proposed that relies on tube smashing and densification, along with delamination and detachment of the tubes, through cracks formation and propagation from the surface to subsurface regions of the film [101].

An additional important outcome from the study performed by Alves et al. [102], is the ability of TiO_2 NTs to induce the formation of a protective P-rich tribo-film with lubricating properties during tribo-electrochemical solicitations. This film may help to explain the open circuit potential (OCP) evolution during sliding, both for conventional and bio-functionalized NTs, as well as the high electrochemical stability observed in both cases when submitted to multiple sliding actions. Beyond the improved electrochemical properties, the formation of this tribo-film is believed to synergistically improve the wear resistance ability of TiO_2 NTs. Conventional NTs showed similar degradation by mechanical wear after sliding tests carried out for 300 and 1800 s, and furthermore, this trend was kept after two-cycle sliding periods undertaken for 1800 s each. As concerns bio-functionalized NTs, a trend of a gradual degradation was observed for sliding tests carried out for 300 and 1800 s, and no significant differences were registered when two-cycle sliding tests were carried out. These results highlight the ability of NTs to withstand multiple cycles of mechanical solicitations, and suggest that the degradation induced by mechanical wear in the beginning of sliding actions, dictate their long-term degradation. This ability of nanotubular films to avoid further mechanical degradation as long mechanical solicitations take place, is believed to be strongly related with the formation of the compact P-rich tribo-film, which grants both protection against corrosion and wear [101, 102].

In summary, anodizing has emerged as a very promising technique to modify the surface features of Titanium-based materials at a micron- to nano-scale level, providing them key functionalities for osseointegrated implant applications. Easily, by varying processing parameters such as the applied anodic voltage, process duration and electrolyte composition, surfaces with different designs may be fabricated through the growth of oxide films with different morphological/topographical and chemical features, which in turns modulate biological and bacterial responses, as well as their corrosion and tribocorrosion performances.

3.3 Plasma electrolytic oxidation (PEO)

3.3.1 Definition of the process and features of the anodic oxide films

Plasma electrolytic oxidation (PEO), also known as micro arc oxidation (MAO), is an electrochemical process of oxidation with which a ceramic oxide coating is generated on the surface of light alloys such as Titanium, Aluminum, or Magnesium (**Figure 4**). The oxide layers created with this process are characterized by a high hardness, adhesion and improved wear and corrosion resistances [104–108]. In this process, the samples are anodically polarized to high voltages exceeding the dielectric breakdown voltage, generating discrete short-living plasma micro-discharges on the surface [104–107]. The process is based on conventional anodizing, with similar configuration, but higher voltages (300–600 V) and current densities (10–50 A/dm^2) [109]. The high voltages facilitate the diffusion of oxygen ions through the discharge channels present in the oxide film to the substrate surface where they react to form the anodic film [110, 111]. Since the development and extinction of the micro-discharges takes place within microseconds (10^{-4}–10^{-5} s), the substrate does not exceed 100–150°C and thus is not subjected to thermal damage [2].

One difference between conventional or hard anodizing coatings and those obtained through PEO technique, is the thickness of the coating. Also, PEO is a high temperature process, which results in the formation of an oxide with a crystalline structure composed of rutile or anatase [112]. The improved wear resistance of the PEO generated layers is attributed to the presence of high temperature oxides and complex compounds, specially rutile, generated during the rapid melting by the micro-discharges and cooling in contact with the refrigerated electrolyte [2]. These oxide layers present excellent bonding adhesion to the substrate, and high hardness.

The PEO layer generated on Titanium alloys usually consists of two crystallographic phases, i.e., anatase and rutile. While anatase might present poor corrosion resistance against certain acids and halide solutions, rutile generally exhibits better protective properties [113]. Also, rutile has higher stability and mechanical properties than anatase, due to its higher hardness and denser morphology. For all these reasons, the presence of rutile in orthopedic implants is desired in order to improve tribological and corrosion properties of the Titanium alloys.

The microstructure of the PEO coating depends on the substrate material, the electrolyte composition and temperature, the dielectric breakdown, and the process parameters [106, 108, 110]. These coatings are characterized by a porous

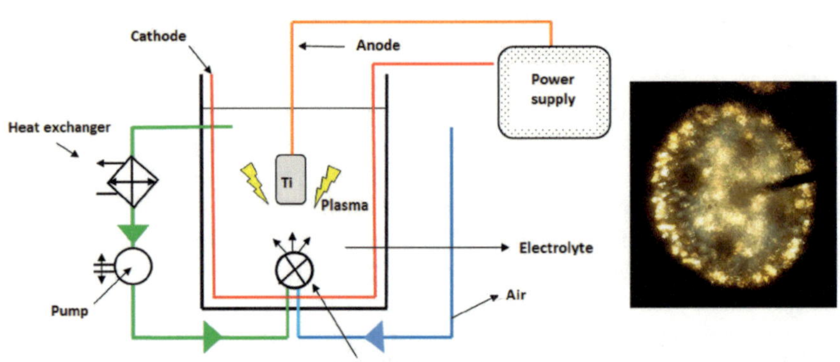

Figure 4.
Schematic illustration of the Plasma Electrolytic Oxidation process and the plasma generated on the surface of a Titanium sample during the process.

microstructure resulting from the gas trapped into the material and the discharge channels created during the process. The higher discharge energy generated by larger spark voltages leads to larger pores in the anodic oxide layer [114]. The oxide layer generated by PEO technique usually consists of two or three layers of different porosity, i.e., an outer porous layer, an intermediate and a denser inner layer. The morphology of the intermediate and outer layers in terms of porosity, thickness, and composition can be tailored controlling the process parameters and the electrolyte composition. The pore diameter in the outer porous layer ranges from 3 to 8 μm in diameter, which results in a relatively high surface roughness [2]. The rough and porous surface of TiO_2 grown by PEO process makes these coatings appropriate for cell adhesion [115].

Furthermore, one of the main reasons for such a good osseointegration of Titanium alloys relay on the mechanical properties of the oxide layer. The elastic modulus of anodic Titanium oxide layers has been reported to be around 40 GPa, which is close to the values for human bones [97]. The low modulus of the oxide layer compared to those of Titanium alloys might be attributed to the porous morphology of this layer.

The use of alkaline electrolytes makes the PEO process sustainable and environmentally friendly. Furthermore, the chemical composition of the electrolyte can be accurately controlled and formulated, in order to control the composition of the resulting oxide layer. For instance, some previous studies on the generation of PEO coatings on Titanium alloys for biomedical implants have incorporated bioactive elements such as phosphorous and calcium on their composition to enhance their osseointegration and favor cell growth [113, 116, 117]. Also, the addition of other elements such as silver, copper, or fluoride to provide the coating with antibacterial ability has also been developed [113, 118–120].

Finally, the easy and rapid operation, and its ecological friendliness make the PEO process a promising technique for industrial applications related to surface modification of Titanium alloys. Furthermore, it is a versatile technique that can be used to coat samples and components of variable size and geometry [121, 122].

3.3.2 Enhancement of Titanium biomedical implants properties by PEO technique

Several investigations have already corroborated the improvement of the tribological properties of Titanium alloys, as well as osseointegration and bactericide properties achieved by means of the PEO technique. In this section, the findings of some studies are presented and briefly described.

The tribological performance of a Ti6Al4V alloy treated by PEO in dry abrasion conditions was investigated by Ceschini et al. [123]. They found the treatment to enhance the behavior of the alloy reducing both the wear extent and the friction coefficient, even under high applied loads up to 35 N. They ascribed this improvement to the high hardness and thickness of the coating, which could appropriately support the applied load protecting the substrate as long as there was a presence of the oxide layer.

The biological response *in vivo* of Titanium alloys treated by PEO technique was confirmed in a study carried out by Ravanetti et al. [124]. They observed the anodic oxide film to promote the early osteoblast adhesion, and the osseointegrative properties *in vivo*. The primary osteogenic response was accelerated during the extensive bone-implant contact after 2 weeks of study.

Alves et al. [125] developed anodic oxide layers on pure Titanium in an electrolyte composed of β-glycerophosphate (β-GP) and calcium acetate (CA). The Ca/P ratio in the resulting films was observed to increase for higher calcium acetate concentrations. In this study, they analyzed the influence of calcium acetate content on the tribocorrosion behavior of the coatings in a sodium chloride solution. The results showed the crystallographic structure of the oxide layer to be affected by the calcium acetate concentration, presenting better tribocorrosion response for higher contents. This was the case of the films with higher Ca/P and rutile/anatase ratios.

Alves et al. [116] employed the PEO technique to develop anodic oxide films on a cp-Ti grade 2 used for dental implants. They evaluated the influence of process parameters on the coating properties, i.e., morphology, roughness, thickness, elemental composition, etc. The corrosion and tribocorrosion tests in artificial saliva (AS) revealed an improvement on the performance of the Titanium after the PEO treatment, in terms of higher corrosion and wear resistances. Also, they stated that the better response was obtained for the samples that were treated under higher current density, which possessed higher layer thickness. Alves et al. [117] continued the previous work by incorporating calcium and phosphorous into the TiO_2 layer of the cp-Ti grade 2 to enhance the osseointegration of the coating. For this aim, they modified the electrolyte composition, and compared the coatings achieved in two electrolytes with two process parameters. Similarly to that observed in the previous work [116], higher process currents and times led to larger pore sizes, and roughness. Also, the electrolyte composition and process time led to variations on the chemical composition and crystalline phases of the coating. The corrosion and tribocorrosion tests in AS revealed the sample with higher rutile/anatase ratio to possess the best performance, with increased mechanical and corrosion resistance. Finally, the cell-material interaction studies carried out with human osteosarcoma cells revealed an improvement on the cell viability/proliferation of the PEO coating compared to the substrate.

Ribeiro et al. [126] employed the PEO technique to develop an oxide layer composed of a mixture of anatase, rutile and amorphous phases. The coating consisted of an outermost nanometric amorphous oxide layer rich in Ca and P. They found this amorphous layer to improve the fibroblast viability and the metabolic activity, as well as the osteoblast adhesion. They observed the osteoblast adhesion to take place preferentially on the amorphous regions with higher Ca content. All the features observed in the layers containing both crystalline and amorphous phases showed a faster osteointegration than those with just crystalline phases.

Oliveira et al. [46] incorporated magnesium together with calcium and phosphorous in the structure of the anodic films, in an electrolyte containing β-GP, CA, and magnesium acetate (MA). They confirmed the addition of magnesium ions to enhance the formation of rutile in the crystalline structure, which also improved the tribocorrosion performance of the developed coating in simulated body fluids (SBF).

Saenz de Viteri et al. [112] developed TiO_2 coatings on Ti6Al4V biomedical alloy used in hip joint stem implants by PEO technique. Apart from Ca and P to generate suitable surface characteristic for cell adhesion and improve the osseointegration, they also incorporated iodine into the composition of the oxide layer to gain antibacterial properties. Iodine has been previously reported to be highly antibacterial, biocompatible and no cytotoxic [127]. However, due to the complexity of its manipulation at room temperature where it sublimes, no previous attempts in its use in PEO coatings has been found in literature. In this work, two process conditions were employed, 5 A/dm^2 for 19 min (PEO-1) and 15 A/dm^2 for 5 min (PEO-2). Similarly to previous results obtained by Alves et al. [116] and Alves et al. [117], the thickness of the coating increased and the pore size and density were higher with higher current density applied. However, the roughness of the oxide was not influenced by this parameter. A typical morphology of the topmost layer of the TiO_2 layer, showing high porosity adequate for cell proliferation is presented in the

Scanning Electron Microscopy (SEM) image of **Figure 5**. The Energy Dispersive Spectroscopy (EDS) and X-Ray Fluorescence (XRF) analysis performed on the samples confirmed the presence of phosphorous, calcium, and iodine in the coatings. Furthermore, the anatase/rutile ratio was quantified by X-Ray Diffraction (XRD), being 21/79 and 46/54 for the PEO-1 and PEO-2, respectively. On the other hand, the tribocorrosion tests in phosphate buffered solution (PBS) showed little influence of sliding on the electrochemical response of the coatings, compared to the Titanium substrate. The OCP lectures were more stable, especially for the PEO-1 coating (see **Figure 5**), and the corrosion resistance of the coatings was not affected after siding. Finally, the antibacterial ability of Titanium against *S. epidermidis* was enhanced by the incorporation of iodine into the coatings.

Marques et al. [118] developed coatings with calcium, phosphorous, silicon, and silver on commercially pure Titanium (Cp-Ti) used in dental implants. The higher presence of rutile in the coatings with higher calcium concentration presented better corrosion properties in AS, also showing antibacterial and biocompatibility properties. They also observed little influence on the coatings performance with the incorporation of silicon and silver into its composition. Finally, the spreading and proliferation of human mesenchymal stem cells on the coatings was optimal, even in the presence of silver nanoparticles that were added to provide antibacterial properties.

Marques et al. [128] evaluated the tribocorrosion behavior of the oxide films containing P, Ca, Si, and Ag, developed in the previous work [118]. Coherently with the former results, the higher amount of Ca led to higher rutile crystalline phase formation on the coating, which was found to possess higher corrosion and tribocorrosion resistances. Also, this coating presented lower material loss, and no detriment was observed in the results with the incorporation of Ag nanoparticles into the coating.

More recently, Chen et al. [129] developed PEO coatings on a Ti-39Nb-6Zr alloy in a KOH solution. They evaluated the corrosion and wear performance of the coated alloy in PBS and studied the biocompatibility by osteoblast cells culturing. The corrosion and wear performance of the alloy was considerably improved after the treatment, showing higher corrosion resistance and lower material loss. The morphology of the oxide layer was found to be beneficial for protein adsorption, and the cell adhesion and osteoblast cell proliferation was also improved by the porous and rough surface.

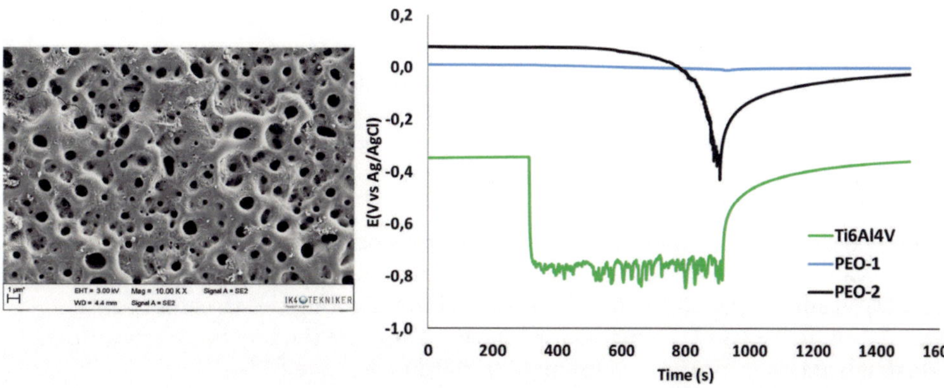

Figure 5.
SEM micrograph of the TiO_2 coating generated by PEO technique on Ti6Al4V presenting adequate morphology for cell proliferation (left) and evolution of the OCP lecture before during and after tribocorrosion for the Ti6Al4V alloy and the two coatings generated by PEO technique [113].

Summarizing, PEO treatment has been found to be a promising surface modification technique for Titanium biomedical alloys. On the one hand, the surface morphology of the resulting TiO_2 layer presents adequate properties for cell proliferations. Furthermore, the electrolyte composition can be tailored modified to successfully incorporate several elements into the oxide composition, to improve different properties. Several researchers used Ca and P to enhance osseointegration, and I or Ag to provide the coating with antibacterial properties. Finally, the corrosion and tribocorrosion properties of Titanium alloys have been found to considerably improved in several biological fluids, e.g., SBF, PBS or AS. The higher enhancement of the TiO_2 layers has been confirmed to occur for greater rutile/anatase ratios.

4. Physical vapor deposition (PVD) technologies

4.1 Definition of PVD technology

Titanium and its alloys exhibit the highest biocompatibility, corrosion resistance, and specific strength (ratio of the tensile strength to density) among different biomaterials [130]. However, they exhibit poor wear resistance and high friction coefficients. Poor abrasive wear resistance results in the formation of wear debris at the implant area, inducing metal ion release, inflammation and pain. Besides this, human life expectancy is steadily growing up, which demands the development of new generation of implants with higher durability. In this sense, significant improvements can be obtained by the application of protective hard coatings by PVD techniques characterized by high hardness and toughness, low elasticity modulus, low friction coefficient and enhanced wear and corrosion resistance in body fluids along with good biocompatibility. PVD techniques are very versatile in terms of the selection of coating material since any type of inorganic material can be deposited.

Physical vapor deposition (PVD) technology refers to a variety of thin film deposition techniques where a solid material (target) is vaporized in high vacuum environment and then transported towards the substrate where it condenses to form a film [131, 132].

PVD processes takes place inside a vacuum chamber where the solid material is evaporated to form a plasma of atoms and molecules that are deposited on a wide range of substrates. As a process that transfers the coating material on a single atom or molecule level, it provides extremely pure and high-performance coatings compared with more conventional techniques. Besides, PVD is an eco-friendly technology in contrast to chemical and galvanic surface treatment methods. It is clean and dry, with no hazardous materials involved, and does not generate chemical waste or water pollution. A wide range of coating materials can be deposited by PVD technologies; including metals, alloys, semiconductors, and ceramics (nitrides, carbides, borides...).

PVD technologies can be divided into different categories regarding the mechanism to vaporize solid material: vacuum evaporation, cathodic arc evaporation, magnetron sputtering and pulsed laser deposition. Among them, cathodic arc evaporation (CAE) and magnetron sputtering (MS) are the most common and industrialized ones for the development of hard protective films.

Cathodic arc evaporation (CAE) is a low voltage, high current plasma discharge that takes place between two metallic electrodes inside a vacuum chamber [133, 134]. The arc discharge current is concentrated at the cathode surface, forming the cathode spots, which are characterized by extremely high current and power densities that produce a localized phase transformation of the solid target (cathode material) to an almost fully ionized deposition plasma with high ion energies. The plasma expands rapidly into the vacuum towards the substrate to form a film. The high number of energetic ions generated in the CAE processes are the main advantage of this technique, which leads to the deposition of extremely dense (suppression of columnar growth) and well-adherent protective films compared with other PVD methods. However, CAE process produces also the so-called "macroparticles" along with the deposition plasma. Macroparticles (molten particles) range in size from less than 1 to about 10 μm in diameter, which adhere to the growing film causing defects, pinholes and rough surface coatings, representing an important disadvantage of CAE technique.

Magnetron sputtering technique relies on ejection (sputtering) of atoms from a solid source (target) by bombardment of gaseous ions from a plasma as shown in **Figure 6** [135–137]. Direct current magnetron sputtering (DCMS) is the basic sputtering process where a DC voltage is applied between the anode (chamber walls) and cathode (target) to ionize the sputtering gas (argon). The positively charged ions (Ar^+) are accelerated towards the cathode leading to an energetic bombardment against the target and ejecting target atoms by momentum transfer mechanism.

The main advantage of magnetron sputtering technology is that almost any material can be sputtered by simply introducing reactive gases such as nitrogen, oxygen or acetylene to form compound materials or by using a radio frequency (RF) power supply to sputtered insulating materials. Besides, there is no macroparticle generation during magnetron sputtering which leads to the development of smoother films compared with CAE technique. The principal drawback of DC magnetron sputtering is that only a small fraction of the sputtered atoms is ionized (<5%) which leads to the deposition of poor adhesion and low-density films characterized by columnar growth.

In order to overcome this problem, high power pulsed magnetron sputtering (HPPMS) was developed in the 1990's as a variation of conventional sputtering by simply changing the power supply used for the generation of plasma discharge

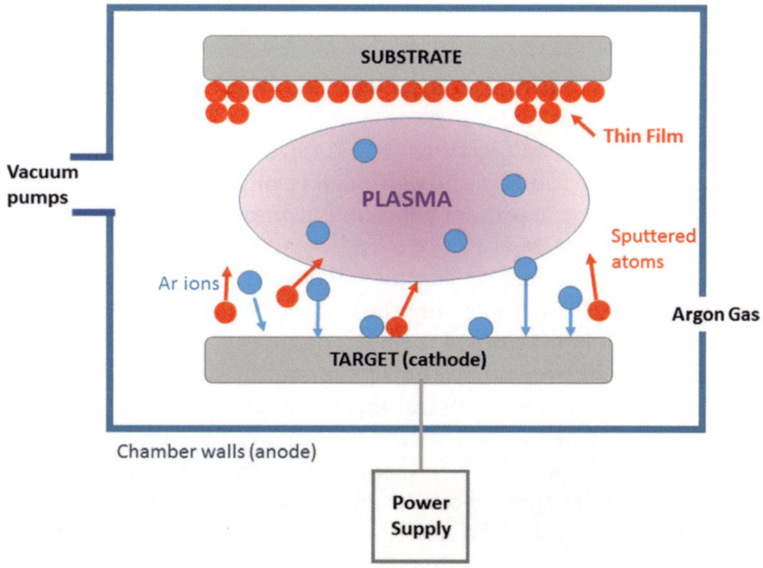

Figure 6.
Schematic illustration of sputtering process.

[138–140]. During HPPMS discharge, the power is applied to the cathode in very short pulses of low duty cycle and frequency which prevents target overheating while increasing peak power density values up to several kW cm^{-2} (two orders of magnitude higher than during dcMS discharge). Such high peak power density values lead to the generation of ultra-dense plasmas, characterized by high ionization degree of sputtered particles. Consequently, thin film growth can be assisted by energetic ion flux bombardment from sputtered material, allowing the modification and densification of film microstructure and development of coatings characterized by excellent adhesion and extremely flat surface [141].

PVD coatings are therefore an excellent alternative to improve surface characteristics of Titanium and its alloys while increasing its durability. Protective biocompatible coatings with high hardness, smooth surface, high corrosion and wear resistance and low friction coefficient can be deposited by PVD techniques [142, 143]. Moreover, they can act as effective barriers to minimize ion release due to tribocorrosion. In addition, antibacterial agents such as silver and/or copper can be incorporated into the coating matrix to try to overcome bacteria colonization problem that occurs on implant devices. PVD process parameters such as pressure, reactive gas flows, power, current and voltage applied during discharge, process temperature...can be tuned to tailor the coating properties and develop an *ad-hoc* coating material for a specific application.

4.2 PVD coatings to improve Titanium and its alloys surface characteristics and performance

Protective coating materials applied by PVD techniques for Titanium and its alloys in biomedical applications generally fall into the following categories:

- diamond-like-carbon (DLC)
- transition metal carbides (MeC_x)
- transition metal nitrides (MeN_X)
- hydroxyapatite (HAP)

DLC is a carbon-based coating composed of diamond (sp^3) and graphite (sp^2) bonds that is commonly applied for the enhancement of Titanium and its alloys performance in biomedical applications [144, 145]. DLC coatings are excellent in both bio- and hemocompatibility, and exhibit very high corrosion resistance and chemical inertness. They can present superior hardness and toughness, besides very low friction coefficient and wear rates depending on the sp^3/sp^2 ratio. The ability to modify the DLC coating surface by doping and changing the compositional variation is an added advantage in improving the coating characteristics [146, 147].

Saenz de Viteri et al. [148] studied the performance of Ti-C-N films deposited on Ti6Al4V alloy by cathodic arc evaporation technique. Five Ti-C-N films were grown using metal Titanium as cathode material and evaporated under different nitrogen (N_2) and acetylene (C_2H_2) flows and arc currents in order to develop Ti-C-N films with different compositions. Tribological study was carried out on Ti6Al4V substrate and Ti-C-N coatings under fretting conditions in the solution of fetal bovine serum (FBS) with sodium azide and EDTA. All coatings reduced the friction coefficient of uncoated Ti6Al4V ($\mu = 0.8$) to values between 0.24 and 0.43 depending on Ti-C-N composition. All coatings could also reduce the wear shown by uncoated Titanium alloy. The coating deposited under lower C_2H_2 flow exhibited the best tribological performance likely due to the higher sp^3/sp^2 fraction which provides the necessary hardness (10 GPa) and the nanocrystalline graphite (nc-G) and amorphous carbon (a-C) structure that confers the lowest friction coefficient and best wear resistance to the coating. Afterwards, Saenz de Viteri et al. [149] did also explore the deposition of silver (Ag) topcoat by magnetron sputtering on Ti-C-N surface to include the antibacterial property on the coating system. Bacterial adhesion tests showed that silver layer has broad-spectrum of anti-adherence activity. However, its low wear resistance makes the layer disappear during rubbing contact, and then their application should be limited to the avoidance of infection during the initial time of coating application.

Bayón et al. [150] investigated the influence of carbon content on the tribocorrosion performance of Titanium DLC (diamond-like-carbon) films deposited on Ti6Al4V alloy. Tribocorrosion analysis was carried out using a ball-on-disc set-up immersed in PBS (phosphate buffer solution) using an alumina ball as a counterpart. Three different deposition processes were accomplished in order to develop three different Ti_DLC coatings by varying some process parameters such as gas flows (nitrogen and acetylene) and arc intensity. DLC1 and DLC3 had same argon and nitrogen content but arc intensity of the process varied from 100 A in case of DLC1 to 75 A in case of DLC3. DLC2 had no nitrogen content on its structure. Corrosion response of uncoated Ti6Al4V and Ti_DLC coatings was very good before wear processes took place. However, high wear rates were measured on Ti6Al4V after tribocorrosion tests due to poor mechanical properties of this material. Ti_DLC films could considerably reduce the friction and wear of substrate material and there was no mechanical contribution to the total material loss due to tribocorrosion, demonstrating the excellent tribological properties of Ti_DLC. Noticeable differences were not detected between the three tested Ti_DLC films.

Braic et al. [151] analysed (Zr, Ti) CN hard coatings deposited by DC magnetron sputtering on Ti6Al4V alloy using Ti and Zr targets sputtered in Argon, nitrogen and methane atmosphere. They compared these films with ZrCN reference coatings. They correlated the composition and crystallographic phase of the films with their mechanical and corrosion properties and surface wettability and cell viability. Two coatings, with different non-metal/metal ratios ($(C + N)/(Zr + Ti) \sim 1.1$ and $(C + N)/(Zr + Ti) \sim 2.6$) were deposited. Smaller grain size and smoother surface was observed for higher non-metal content coating. Contact angle measurements and corrosion tests demonstrated that the deposited coatings were hydrophobic and had improved corrosion resistance as compared with the Ti6Al4V substrate. The coating with higher non-metal/metal ratio exhibited the best performance and higher cell viability. The quaternary (Ti, Zr) CN films were found to have better characteristics than ZrCN reference films being better candidates to protect Ti6Al4V in orthopedic implants.

Transition metal nitrides (MeN_x) are generally used as protective hard coatings due to their high hardness and toughness, low elasticity modulus, enhance wear and corrosion resistance and excellent chemical stability. Among them, tantalum nitride (TaN) is particularly interesting for Titanium protection in biomedical applications owing to its outstanding biocompatibility. Tantalum metal is one of the most chemically inert and biocompatible material, showing an outstanding corrosion performance in many corrosive environments, even comparable to that of noble metals [152].

Mendizabal et al. [153] presented an investigation about TaN films deposited by HPPMS technique on pure Titanium (Ti-cp) grade 2 to be applied on biomedical implants. They investigated three different TaN films; i.e., monolayer TaN and two multilayer TaN films characterized by different bilayer periods were developed by alternatively switching two different HPPMS pulses within one overall deposition process. The microstructural analysis of the TaN films (**Figure 7**) revealed

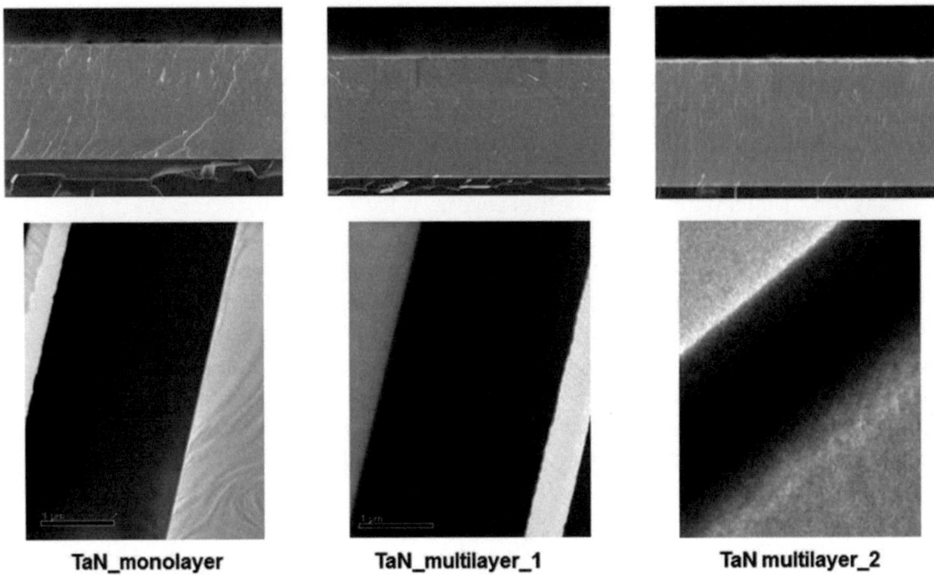

Figure 7.
Cross-sectional SEM and TEM micrographs of monolayer and multilayer TaN films [153].

extremely dense microstructures and suppression of columnar growth for all deposited samples confirming the importance of highly ionized plasmas developed during HPPMS discharges on the densification of the films.

Since the natural physiological environment contains not only inorganic species but also organic molecules such as serum proteins [154], Phosphate Buffered Solution (PBS) plus 1 g of albumin was chosen as experimental electrolyte to perform corrosion and tribocorrosion testing. The corrosion resistance of Ti-cp and TaN coatings before sliding was incredibly high (~MΩ) and increased with time. TaN coatings exhibit similar electrochemical responses and slightly enhanced the Ti-cp resistance. The corrosion resistance of Ti-cp during sliding decreased up to 1 kΩ and TaN coatings exhibited one order of magnitude higher resistance values. The friction coefficient was reduced from 0.58 to 0.25 by all TaN coatings. The wear rate of Ti-cp was considerably reduced, and total material loss caused by tribocorrosion reduced up to 96% by best-performing TaN multilayer film.

Huang et al. [155] studied the incorporation of Ag on a TaN coating deposited on pure Titanium by co-sputtering of Ag and Ta targets in Ar/N_2 atmosphere in order to evaluate the antibacterial property of the system. They also investigated the biocompatibility of different TaN-Ag coatings grown with different Ag contents. The pure TaN showed a dense columnar structure with smooth surface and possessed the highest contact angle showing hydrophobic characteristics. The incorporation of Ag on the TaN matrix lowered the contact angle of the system. The TaN-Ag composite coatings with the highest Ag content (21.4 at.%) showed the most significant short-term antibacterial effect. All TaN_Ag films met the requirements in terms of cell viability independently from Ag content.

Hydroxyapatite (HAP) is the most common phase among calcium phosphate (CaP) ceramics studied for biomedical applications due to its similarity to natural bone and owing to their characteristics such as high biocompatibility, osteoconductivity, chemical stability at a neutral pH, and osseointegration [156]. HAP is, therefore, frequently considered as an adequate coating on Titanium orthopedic implants, which has been proven to promote and accelerate the osseointegration of an implant device into the body [157]. However, a major concern regarding the application of HAP coatings is the low mechanical properties and high rate of decomposition of this material, which hinders its commercial application in high durability demanding devices. In order to overcome this problem, Vranceanu et al. [158] studied the addition of silicon carbide (SiC) into hydroxyapatite films deposited by RF-magnetron sputtering technique on Ti6Al4V for orthopedic applications. Previous results showed the improvement of wear and corrosion resistance of HAP films due to Sic incorporation in Titanium for dental implants [159, 160]. They studied the performance of base HAP and three HAP + SiC with different SiC contents by varying power applied to SiC sputtering target. They found that the SiC addition into the HAP matrix considerably improved the adhesion of the coating to the Ti6Al4V substrate and corrosion resistance compared to both base HAP film and uncoated Ti6Al4V. The biocompatibility tests demonstrated better cell proliferation and viability of all coatings compared with uncoated Ti6Al4V.

Summarizing, PVD technology is very versatile in terms of the available type of coating materials and properties that can be selected and tailored to enhance the performance and durability of Titanium and its alloys. Particularly,

- DLC and carbon-based films can strongly reduce the friction coefficient and wear rate of Titanium and its alloys in simulated body fluids.
- TaN films can considerably reduce the total material loss due to tribocorrosion observed for uncoated pure Titanium.
- The incorporation of silver into any of these coating matrixes can add the antibacterial property to the films without affecting their biocompatibility.
- Hydroxyapatite (HAP) films with enhanced adhesion and corrosion resistance due to SiC incorporation can be deposited by magnetron sputtering. Besides, HAP + SiC exhibited better cell proliferation and viability than uncoated Ti6Al4V.

The broad type of coating materials that can be deposited by PVD techniques, along with the possibility to tune their properties by fine adjustment of process parameters, presents these technologies as excellent candidates to enhance the durability of Titanium and its alloys for biomedical applications.

5. Conclusions

In summary, among the various electrochemical surface modification technologies nowadays available to enhance Titanium-based materials features for biomedical applications, anodizing and PEO techniques are among the most promising ones, for the following reasons:

- By varying processing parameters such as the process voltage, duration and electrolyte composition, surfaces with different designs may be fabricated by changing their topographical/morphological features, as well as their chemical composition.
- The electrolyte composition can be tailored to successfully incorporate several elements into the oxide film composition, to improve different properties. The enrichment of the oxide film structure with bioactive and antimicrobial

agents may induce to the production of new bio-selective functional surfaces for osseointegrated Titanium implants, by improving osseointegration and simultaneously avoiding infection.

- The growth of an oxide film on Titanium surfaces may improve considerably their corrosion and tribocorrosion properties in simulated body fluids, such as artificial saliva. In the case of PEO, which generally induces to the growth of an oxide film with a crystalline structure, the improved corrosion and tribocorrosion performances of TiO_2 films have been confirmed to occur for greater rutile/anatase ratios.

Furthermore, PVD technologies are very versatile and they are characterized for the development of well-adherent, hard, dense and high durability films. A broad range of coating materials can be grown by PVD, made of different compounds or in multilayer structures under low temperature deposition processes. Fine-tuning of process parameters enables the development of completely different surface features making possible the deposition of *ad-hoc* protective coatings. It must be highlighted:

- Very low friction coefficient and wear resistance carbon-based films are deposited by PVD techniques to protect Titanium from mechanical degradation;

- Transition metal nitride films with high corrosion and wear resistance are grown by PVD techniques, which considerably improve tribocorrosion performance of Titanium;

- The incorporation of silver into any of these coating matrixes can add the antibacterial property to the films without affecting their biocompatibility;

- Mechanical properties of hydroxyapatite films can be enhanced by SiC incorporation using PVD techniques, maintaining the excellent osseointegration of HAP while increasing its durability.

To conclude, this chapter is focused on promising electrochemical and physical technologies to modify the surface features of Titanium-based materials aiming to improve their functionalities for biomedical implant applications. In general, the implant surface requirements are defined based on its final application. Therefore, the most adequate technique for surface treatment must selected considering the required surface features of the implant, namely topography, morphology, thickness, chemistry, hardness, as well as its response regarding corrosion and tribocorrosion processes. Based on this approach, implants with single or multiple functionalities may be designed, namely osseointegrable implants with antimicrobial properties, simultaneously displaying improved corrosion and tribocorrosion responses.

Acknowledgements

This work was performed with the financial support of the EMAITEK and ELKARTEK 2018 (FRONTIERS IV project, KK-2018/00108) research programs from the Basque Government, and the FUNCOAT project (CSD2008-00023) from the CONSOLIDER INGENIA-2010 program financed by the Spanish National Department of the Education and Science.

Author details

Elena Fuentes[1*], Sofia Alves[1], Ainara López-Ortega[1], Lucía Mendizabal[1] and Virginia Sáenz de Viteri[2]

1 IK4-Tekniker, Eibar, Spain

2 Mugape, Mallabia, Spain

*Address all correspondence to: elena.fuentes@tekniker.es

References

[1] Neoh KG, Hu X, Zheng D, Kang ET. Balancing osteoblast functions and bacterial adhesion on functionalized titanium surfaces. Biomaterials. 2012;**33**:2813-2822. DOI: 10.1016/j.biomaterials.2012.01.018

[2] Sáenz de Viteri V, Fuentes E. Titanium and titanium alloys as biomaterials. In: Gegner J, editor. Tribology. Fundamentals and Advancements. IntechOpen; 2013. pp. 155-181. ISBN: 978-953-51-1135-1. DOI: 10.5772/55860

[3] Valles G, Perez C, Bore A, Martin-Saavedra F, Saldana L, Vilaboa N. Simvastatin prevents the induction of interleukin-6 gene expression by titanium particles in human osteoblastic cells. Acta Biomaterialia. 2013;**9**:4916-4925. DOI: 10.1016/j.actbio.2012.08.027

[4] Saldaña L, Vilaboa N. Effects of micrometric titanium particles on osteoblast attachment and cytoskeleton architecture. Acta Biomaterialia. 2010;**6**:1649-1660. DOI: 10.1016/j.actbio.2009.10.033

[5] Khan SN, Ramachandran M, Senthil Kumar S, Krishnan V, Sundaram R. Osseointegration and more—A review of literature. Indian Journal of Dentistry. 2012;**3**:72-76. DOI: 10.1016/j.ijd.2012.03.012

[6] Stanford CM. Surface modification of biomedical and dental implants and the processes of inflammation, wound healing and bone formation. International Journal of Molecular Sciences. 2010;**11**:354-369. DOI: 10.3390/ijms11010354

[7] Souza JCM, Barbosa SL, Ariza E, Celis JP, Rocha LA. Simultaneous degradation by corrosion and wear of titanium in artificial saliva containing fluorides. Wear. 2012;**292-293**:82-88. DOI: 10.1016/j.wear.2012.05.030

[8] Imai Y, Yokoyama A, Yamamoto S, Obata T, Iizuka T, Kohgo T, et al. Peri-implant tissue after osseointegration in diabetes in rat maxilla. Journal of Oral Biosciences. 2006;**48**:54-61. DOI: 10.2330/joralbiosci.48.54

[9] Subbiahdoss G, Kuijer R, Grijpma DW, van der Mei HC, Busscher HJ. Microbial biofilm growth vs. tissue integration: "The race for the surface" experimentally studied. Acta Biomaterialia. 2009;**5**:1399-1404. DOI: doi.org/10.1016/j.actbio.2008.12.011

[10] Mathew MT, Srinivasa Pai P, Pourzal R, Fischer A, Wimmer MA. Significance of tribocorrosion in biomedical applications: Overview and current status. Advances in tribology. 2009:12. Article ID 250986. DOI: 10.1155/2009/250986

[11] Zhu MH, Cai ZB, Li W, Yu HY, Zhou ZR. Fretting in prosthetic devices related to human body. Tribology International. 2009;**42**:1360-1364. DOI: 10.1016/j.triboint.2009.04.007

[12] Fraczek T, Olejnik M, Tokarz A. Evaluations of plasma nitriding efficiency of titanium alloys for medical applications. Metalurgija. 2009;**48**(2):83-86. DOI: 10.1002/3527602119.ch16

[13] Kustas FM, Misra MS. Friction and wear of titanium alloys. In: Henry SD, editor. Friction, Lubrication and Wear Technology. Vol. 18. United States of America: ASM International; 1992. pp. 1585-1598

[14] Freese H, Volas MG, Wood JR. In: Brunette DM, Tengvall P, Textor M, Thomsen P, editors. Titanium in Medicine. United States of America: Springer; 2001. p. 25-51

[15] Vieira AC, Ribeiro AR, Rocha LA, Celis JP. Influence of pH and corrosion

inhibitors on the tribocorrosion of titanium in artificial saliva. Wear. 2006;**261**:994-1001. DOI: 10.1016/j.wear.2006.03.031

[16] Long M, Rack HJ. Friction and surface behavior of selected titanium alloys during reciprocating-sliding motion. Wear. 2001;**249**:157-167. DOI: 10.1016/S0043-1648(01)00517-8

[17] Yerokhin AL, Niea X, Leyland A, Matthews A. Characterization of oxide films produced by plasma electrolytic oxidation of a Ti-6Al-4V alloy. Surface & Coating Technology. 2000;**130**:195-206. DOI: 10.1016/S0257-8972(00)00719-2

[18] Manivasagam G, Dhinasekaran D, Rajamanickam A. Biomedical implants: Corrosion and its prevention—A review recent patents on corrosion science. 2010;**2**:40-54. DOI: 10.2174/1877610801002010040

[19] Adya N, Alam M, Ravindranath T, Mubeen A, Saluja B. Corrosion in titanium dental implants: Literature review. The Journal of Indian Prosthodontic Society. 2005;**5**:126-131

[20] Celis JP, Ponthiaux P, editors. Testing Tribocorrosion of Passivating Materials Supporting Research and Industrial Innovation: Handbook. Maney Publishing; 2012. ISBN: 978-1-907975-20-2

[21] Landolt D. Corrosion and Surface Chemistry of Metals. Lausanne, Switzerland: EPFL Press; 2007. ISBN: 9780849382338

[22] Virtanen S, Milošev I, Gomez-Barrena E, Trebše R, Salo J, Konttinen YT. Special modes of corrosion under physiological and simulated physiological conditions. Acta Biomaterialia. 2008;**4**:468-476. DOI: 10.1016/j.actbio.2007.12.003

[23] Barão VAR, Mathew MT, Assunção WG, Yuan JCC, Wimmer MA, Sukotjo C. Stability of cp-Ti and Ti-6Al-4V alloy for dental implants as a function of saliva pH—An electrochemical study. Clinical Oral Implants Research. 2012;**23**:1055-1062. DOI: 10.1111/j.1600-0501.2011.02265.x

[24] Takemoto S, Hattori M, Yoshinari M, Kawada E, Oda Y. Corrosion behavior and surface characterization of titanium in solution containing fluoride and albumin. Biomaterials. 2005;**26**: 829-837. DOI: 10.1016/j.biomaterials.2004.03.025

[25] Souza JCM, Ponthiaux P, Henriques M, Oliveira R, Teughels W, Celis J-P, et al. Corrosion behaviour of titanium in the presence of *Streptococcus mutans*. Journal of Dentistry. 2013;**41**:528-534. DOI: 10.1016/j.jdent.2013.03.008

[26] Cadosch D, Al-Mushaiqri MS, Gautschi OP, Meagher J, Simmen HP, Filgueira L. Biocorrosion and uptake of titanium by human osteoclasts. Journal of Biomedical Materials Research Part A. 2010;**95**:1004-1010. DOI: 10.1002/jbm.a.32914

[27] Cruz HV, Souza JCM, Henriques M, Rocha LA. Tribocorrosion and bio-tribocorrosion in the oral environment: The case of dental implants. In: Davim JP, editor. Biomedical Tribology. Portugal: Nova Science Publishers; 2011. pp. 1-30. ISBN: 978-1-61470-056-2

[28] Black J. Othopaedic Biomaterials in Research and Practice. New York: Churchill Livingstone; 1988. ISBN: 9781138074866

[29] Jacobs JJ, Gilbert JL, Urban RM. Corrosion of metal orthopaedic implants. Journal of Bone and Joint Surgery: American Volume. 1998;**80**:268-282. DOI: 10.1097/00004623-199802000-0001

[30] Damborenea JJ, Vázquez AJ. Consejo Superior de Investigaciones Científicas.

Ciencia e ingeniería de la superficie de los materiales metálicos. Raycar. Spain; 2000. ISBN: 978-84-00-07920-8

[31] Baboian R, editor. ASTM Corrosion Tests and Standards. Application and Interpretation. 2nd ed. Baltimore: ASTM International; 2006. ISBN: 978-0-8031-2098-2

[32] Shih H, editor. Corrosion Resistance. United Kingdom: In Tech; 2012. DOI: 10.5772/1844. ISBN: 978-953-51-0467-4

[33] Landolt D, Mischler S, editors. Tribocorrosion of Passive Metals and Coatings. Cambridge: Woodhead Publishing; 2011. ISBN: 978-1-84569-966-6

[34] Landolt D. Electrochemical and materials aspects of tribocorrosion systems. Journal of Physics D. 2006;**39**:3121-3127. DOI: 10.1088/0022-3727/39/15/S01

[35] Mischler S, Rosset EA, Landolt D. Effect of corrosion on the wear behaviour of passivating metals in aqueous solutions. Tribology Series. 1993;**25**:245-253. DOI: 10.1016/S0167-8922(08)70380-8

[36] López-Ortega A, Arana JL, Bayón R. Tribocorrosion of passive materials: A review on test procedures and standards. International Journal of Corrosion. 2018:24 p. Article ID: 7345346. DOI: 10.1155/2018/7345346

[37] Song HJ, Kim MK, Jung GC, Vang MS, Park YJ. The effects of spark anodizing treatment of pure titanium metals and titanium alloys on corrosion characteristics. Surface and Coatings Technology. 2007;**201**:8738-8745. DOI: 10.1016/j.surfcoat.2006.11.022

[38] Cobelli N, Scharf B, Crisi GM, Hardin J, Santambrogio L. Mediators of the inflammatory response to joint replacement devices. Nature Reviews Rheumatology. 2011;7:600-608. DOI: 10.1038/nrrheum.2011.128

[39] Saidin S, Abdul Kadir MR, Sulaiman E, Abu Kasim NH. Effects of different implant—Abutment connections on micromotion and stress distribution: Prediction of microgap formation. Journal of Dentistry. 2012;**40**:467-474. DOI: 10.1016/j.jdent.2012.02.009

[40] do Nascimento C, Barbosa RES, Issa JPM, Watanabe E, Ito IY, Albuquerque Junior RF. Bacterial leakage along the implant—Abutment interface of premachined or cast components. International Journal of Oral and Maxillofacial Surgery. 2008;**37**:177-180. DOI: 10.1016/j.ijom.2007.07.026

[41] Alves SA, Beline T, Barão VAR, Sukotjo C, Mathew MT, Rocha LA, et al. Chapter 3—Degradation of titanium-based implants. In: Sousa JCM, Hotza D, Henriques B, Boccaccini AR, editors. Nanostructured Biomaterials for Cranio-maxillofacial and Oral Applications. 1st ed. Nederland: Elsevier; 2018. pp. 41-46. DOI: 10.1016/B978-0-12-814621-7.00003-2

[42] Cruz HV, Souza JCM, Henriques M, Rocha LA. Chapter 1—Tribocorrosion and bio-tribocorrosion in the oral environment: The case of dental implants. In: Davim JP, editor. Biomedical Tribology. New York: Nova Science Publishers Inc.; 2011. pp. 1-30

[43] Jacobs JJ, Hallab NJ, Urban RM, Wimmer MA. Wear particles. The Journal of Bone and Joint Surgery: American Volume. 2006;**88**(2):99-102. DOI: 10.2106/JBJS.F.00102

[44] Kieswetter K, Schwartz Z, Dean DD, Boyan BD. The role of implant surface characteristic in the healing of bone. Critical Reviews in Oral Biology & Medicine. 1996;7:329-345

[45] Keller JC, Stanford CM, Wightman JP, Draughn RA, Zaharias R. Characterization of titanium implant surfaces. III. Journal of Biomedical Materials Research. 1994;**28**:939-946. DOI: 10.1002/jbm.820280813

[46] Oliveira FG, Ribeiro AR, Perez G, Archanjo BS, Gouvea CP, Araújo JR, et al. Understanding growth mechanisms and tribocorrosion behaviour of porous TiO_2 anodic films containing calcium, phosphorous and magnesium. Applied Surface Science. 2015;**341**:1-12. DOI: 10.1016/j.apsusc.2015.02.163

[47] Kim KH, Ramaswamy N. Electrochemical surface modification of titanium in dentistry. Dental Materials Journal. 2009;**28**:20-36. DOI: 10.4012/dmj.28.20

[48] Li LH, Kong YM, Kim HW, Kim YW, Kim HE, Heo SJ, et al. Improved biological performance of Ti implants due to surface modification by micro-arc oxidation. Biomaterials. 2004;**25**:2867-2875. DOI: 10.1016/j.biomaterials.2003.09.048

[49] Yang WE, Hsu ML, Lin MC, Chen ZH, Chen LK, Huang HH. Nano/submicron-scale TiO_2 network on titanium surface for dental implant application. Journal of Alloys and Compounds. 2009;**479**:642-647. DOI: 10.1016/j.jallcom.2009.01.021

[50] Yao C, Webster TJ. Anodization: a promising nano-modification technique of titanium implants for orthopedic applications. Journal of nanoscience and nanotechnology. 2006;**6**:2682-2692. DOI: 10.1166/jnn.2006.447

[51] Roy P, Berger S, Schmuki P. TiO_2 nanotubes: Synthesis and applications. Angewandte Chemie International Edition. 2011;**50**:2904-2939. DOI: 10.1002/anie.201001374

[52] Rao BM, Torabi A, Varghese OK. Anodically grown functional oxide nanotubes and applications. MRS Communications. 2016;**6**:375-396. DOI: 10/1557/mrc.2016.46

[53] Sul YT, Johansson CB, Petronis S, Krozer A, Jeong Y, Wennerberg A, et al. Characteristics of the surface oxides on turned and electrochemically oxidized pure titanium implants up to dielectric breakdown: The oxide thickness, micropore configurations, surface roughness, crystal structure and chemical composition. Biomaterials. 2002;**23**:491-501. DOI: 10.1016/S0142-9612(01)00131-4

[54] Hu Y, Cai K, Luo Z, Xu D, Xie D, Huang Y, et al. TiO_2 nanotubes as drug nanoreservoirs for the regulation of mobility and differentiation of mesenchymal stem cells. Acta Biomaterialia. 2012;**8**:439-448. DOI: 10.1016/j.actbio.2011.10.021

[55] Demetrescu I, Pirvu C, Mitran V. Effect of nano-topographical features of Ti/TiO_2 electrode surface on cell response and electrochemical stability in artificial saliva. Bioelectrochemistry. 2010;**79**:122-129. DOI: 10.1016/j.bioelechem.2010.02.001

[56] Yao C, Slamovich EB, Webster TJ. Enhanced osteoblast functions on anodized titanium with nanotube-like structures. Journal of Biomedical Materials Research. Part A. 2008;**85**:57-66. DOI: 10.1002/jbm.a.31551

[57] Indira K, Mudali UK, Nishimura T, Rajendran N. A review on TiO_2 nanotubes: Influence of anodization parameters, formation mechanism, properties, corrosion behavior, and biomedical applications. Journal of Bio- and Tribo-Corrosion. 2015;**1**:1-28. DOI: 10.1007/s40735-015-0024-x

[58] Ge M, Li Q, Cao C, Huang J, Li S, Zhang S, et al. One-dimensional TiO_2 nanotube photocatalysts for solar water

splitting. Advanced Science. 2016;**4**:1-31. DOI: 10.1002/advs.201600152

[59] Regonini D, Bowen CR, Jaroenworaluck A, Stevens R. A review of growth mechanism, structure and crystallinity of anodized TiO_2 nanotubes. Materials Science and Engineering: R: Reports. 2013;**74**:377-406. DOI: 10.1016/j.mser.2013.10.001

[60] Macak JM, Tsuchiya H, Ghicov A, Yasuda K, Hahn R, Bauer S, et al. TiO_2 nanotubes: Self-organized electrochemical formation, properties and applications. Current Opinion in Solid State and Materials Science. 2007;**11**:3-18. DOI: 10.1016/j.cossms.2007.08.004

[61] Brammer KS, Frandsen CJ, Jin S. TiO_2 nanotubes for bone regeneration. Trends in Biotechnology. 2012;**30**:315-322. DOI: 10.1016/j.tibtech.2012.02.005

[62] Macak JM, Sirotna K, Schmuki P. Self-organized porous titanium oxide prepared in Na_2SO_4/NaF electrolytes. Electrochimica Acta. 2005;**50**:3679-3684. DOI: 10.1016/j.electacta.2005.01.014

[63] Shankar K, Mor GK, Prakasam HE, Yoriya S, Paulose M, Varghese OK, et al. Highly-ordered TiO_2 nanotube arrays up to 220 μm in length: use in water photoelectrolysis and dye-sensitized solar cells. Nanotechnology. 2007;**18**:1-11. DOI: 10.1088/0957-4484/18/6/065707

[64] Han SC, Doh JM, Yoon JK, Kim GH, Byun JY, Han SH, et al. Highly ordered self-organized TiO_2 nanotube arrays prepared by a multi-step anodic oxidation process. Metals and Materials International. 2009;**15**:493-499. DOI: 10.1007/s12540-009-0493-x

[65] Sulka GD, Kapusta-Kołodziej J, Brzózka A, Jaskuła M. Fabrication of nanoporous TiO_2 by electrochemical anodization. Electrochimica Acta. 2010;**55**:4359-4367. DOI: 10.1016/j.electacta.2009.12.053

[66] Ali G, Chen C, Yoo SH, Kum JM, Cho SO. Fabrication of complete titania nanoporous structures via electrochemical anodization of Ti. Nanoscale Research Letters. 2011;**6**:1-10. DOI: 10.1186/1556-276X-6-332

[67] Peighambardoust NS, Nasirpouri F. Manipulating morphology, pore geometry and ordering degree of TiO_2 nanotube arrays by anodic oxidation. Surface and Coatings Technology. 2013;**235**:727-734. DOI: 10.1016/j.surfcoat.2013.08.058

[68] Gulati K, Maher S, Findlay DM, Losic D. Titania nanotubes for orchestrating osteogenesis at the bone—Implant interface. Nanomedicine. 2016;**11**:1847-1864. DOI: doi.org/10.2217/nnm-2016-0169

[69] Anselme K, Davidson P, Popa AM, Giazzon M, Liley M, Ploux L. The interaction of cells and bacteria with surfaces structured at the nanometre scale. Acta Biomaterialia. 2010;**6**:3824-3846. DOI: 10.1016/j.actbio.2010.04.001

[70] Zhao L, Mei S, Chu PK, Zhang Y, Wu Z. The influence of hierarchical hybrid micro/nano-textured titanium surface with titania nanotubes on osteoblast functions. Biomaterials. 2010;**31**:5072-5082. DOI: 10.1016/j.biomaterials.2010.03.014

[71] Zhao L, Liu L, Wu Z, Zhang Y, Chu PK. Effects of micropitted/nanotubular titania topographies on bone mesenchymal stem cell osteogenic differentiation. Biomaterials. 2012;**33**:2629-2641. DOI: 10.1016/j.biomaterials.2011.12.024

[72] Zhang W, Wang G, Liu Y, Zhao X, Zou D, Zhu C, et al. The

synergistic effect of hierarchical micro/nano-topography and bioactive ions for enhanced osseointegration. Biomaterials. 2013;**34**:3184-3195. DOI: 10.1016/j.biomaterials.2013.01.008

[73] Popat KC, Leoni L, Grimes CA, Desai TA. Influence of engineered titania nanotubular surfaces on bone cells. Biomaterials. 2007;**28**:3188-3197. DOI: 10.1016/j.biomaterials.2007.03.020

[74] Wang N, Li H, Lü W, Li J, Wang J, Zhang Z, et al. Effects of TiO_2 nanotubes with different diameters on gene expression and osseointegration of implants in minipigs. Biomaterials. 2011;**32**:6900-6911. DOI: 10.1016/j.biomaterials.2011.06.023

[75] Salou L, Hoornaert A, Louarn G, Layrolle P. Enhanced osseointegration of titanium implants with nanostructured surfaces: An experimental study in rabbits. Acta Biomaterialia. 2015;**11**:494-502. DOI: 10.1016/j.actbio.2014.10.017

[76] Brammer KS, Oh S, Gallagher JO, Jin S. Enhanced cellular mobility guided by TiO_2 nanotube surfaces. Nano letters. 2008;**8**:786-793. DOI: 10.1021/nl072572o

[77] Park J, Bauer S, Schlegel KA, Neukam FW, von der Mark K, Schmuki P. TiO_2 nanotube surfaces: 15 nm—An optimal length scale of surface topography for cell adhesion and differentiation. Small. 2009;**5**:666-671. DOI: 10.1002/smll.200801476

[78] Shi X, Xu Q, Tian A, Tian Y, Xue X, Sun H, et al. Antibacterial activities of TiO_2 nanotubes on *Porphyromonas gingivalis*. RSC Advances. 2015;**5**:34237-34242. DOI: 10.1039/c5ra00804b

[79] Li H, Cui Q, Feng B, Wang J, Lu X, Weng J. Antibacterial activity of TiO_2 nanotubes: Influence of crystal phase, morphology and Ag deposition. Applied Surface Science. 2013;**284**:179-183. DOI: 10.1016/j.apsusc.2013.07.076

[80] Ercan B, Taylor E, Alpaslan E, Webster TJ. Diameter of titanium nanotubes influences anti-bacterial efficacy. Nanotechnology. 2011;**22**:1-11. DOI: 10.1088/0957-4484/22/29/295102

[81] Park J, Bauer S, von der Mark K, Schmuki P. Nanosize and vitality: TiO_2 nanotube diameter directs cell fate. Nano letters. 2007;**7**:1686-16891. DOI: 10.1021/nl070678d

[82] Bauer S, Park J, Faltenbacher J, Berger S, von der Mark K, Schmuki P. Size selective behavior of mesenchymal stem cells on ZrO_2 and TiO_2 nanotube arrays. Integrative Biology. 2009;**1**:525-532. DOI: 10.1039/b908196h

[83] Oh S, Brammer KS, Li YSJ, Teng D, Engler AJ, Chien S, et al. Stem cell fate dictated solely by altered nanotube dimension. Proceedings of the National Academy of Sciences. 2009;**106**:2130-2135. DOI: 10.1073/pnas.0813200106

[84] Grotberg J, Hamlekhan A, Butt A, Patel S, Royhman D, Shokuhfar T, et al. Thermally oxidized titania nanotubes enhance the corrosion resistance of Ti6Al4V. Materials Science and Engineering: C. 2016;**59**:677-689. DOI: 10.1016/j.msec.2015.10.056

[85] Yu WQ, Qiu J, Xu L, Zhang FQ. Corrosion behaviors of TiO_2 nanotube layers on titanium in Hank's solution. Biomedical Materials. 2009;**4**:1-6. DOI: 10.1088/1748-6041/4/6/065012

[86] Wu S, Weng Z, Liu X, Yeung KWK, Chu P. Functionalized TiO_2 based nanomaterials for biomedical applications. Advanced Functional Materials. 2014;**24**:1-18. DOI: 10.1002/adfm.201400706

[87] Hamlekhan A, Sinha-Ray S, Takoudis C, Mathew MT, Sukotjo C, Yarin AL, et al. Fabrication of drug

eluting implants: Study of drug release mechanism from titanium dioxide nanotubes. Journal of Physics D: Applied Physics. 2015;**48**:1-11. DOI: 10.1088/0022-3727/48/27/275401

[88] Popat KC, Eltgroth M, LaTempa TJ, Grimes CA, Desai TA. Titania nanotubes: A novel platform for drug-eluting coatings for medical implants? Small. 2007;**3**:1878-1881. DOI: 10.1002/smll.200700412

[89] Alves SA, Patel SB, Sukotjo C, Mathew MT, Filho PN, Celis J-P, et al. Synthesis of calcium-phosphorous doped TiO_2 nanotubes by anodization and reverse polarization: A promising strategy for an efficient biofunctional implant surface. Applied Surface Science. 2017;**399**:682-701. DOI: 10.1016/j.apsusc.2016.12.105

[90] Alves SA , Rossi AL, Ribeiro AR, Gemini-Piperni S, Silva RC, Saraiva AM, et al. TiO_2 nanotubes enriched with calcium, phosphorous and zinc: Promising bio-selective functional surfaces for osseointegrated titanium implants. RSC Advances. 2017;7: 49720-49738. DOI: 10.1039/C7RA08263K

[91] Crawford GA, Chawla N, Houston J. Nanomechanics of biocompatible TiO_2 nanotubes by interfacial force microscopy (IFM). Journal of the Mechanical Behavior of Biomedical Materials. 2009;**2**:580-587. DOI: 10.1016/j.jmbbm.2008.10.004

[92] Crawford GA, Chawla N, Das K, Bose S, Bandyopadhyay A. Microstructure and deformation behavior of biocompatible TiO_2 nanotubes on titanium substrate. Acta Biomaterialia. 2007;**3**:359-367. DOI: 10.1016/j.actbio.2006.08.004

[93] Xu YN, Liu MN, Wang MC, Oloyede A, Bell JM, Yan C. Nanoindentation study of the mechanical behavior of TiO_2 nanotube arrays. Journal of Applied Physics. 2015;**118**:1-7. DOI: 10.1063/1.4932213

[94] Rho JY, Kuhn-Spearing L, Zioupos P. Mechanical properties and the hierarchical structure of bone. Medical Engineering & Physics. 1998;**20**:92-102. DOI: 10.1016/S1350-4533(98)00007-1

[95] Wang X, Xu S, Zhou S, Xu W, Leary M, Choong P, et al. Topological design and additive manufacturing of porous metals for bone scaffolds and orthopaedic implants: A review. Biomaterials. 2016;**83**:127-141. DOI: 10.1016/j.biomaterials.2016.01.012

[96] Rho JY, Tsui TY, Pharr GM. Elastic properties of human cortical and trabecular lamellar bone measured by nanoindentation. Biomaterials. 1997;**18**:1325-1330. DOI: 10.1016/s0142-9612(97)00073-2

[97] Soares P, Mikowski A, Lepienski CM, Santos E, Soares GA, Kuromoto NK. Hardness and elastic modulus of TiO_2 anodic films measured by instrumented indentation. Journal of Biomedical Materials Research Part B: Applied Biomaterials. 2008;**84**:524-530

[98] Yu D, Zhu X, Xu Z, Zhong X, Gui Q, Song Y, et al. Facile method to enhance the adhesion of TiO_2 nanotube arrays to Ti substrate. ACS Applied Materials & Interfaces. 2014;**6**: 8001-8005. DOI: 10.1021/am5015716

[99] Ouyang HM, Fei GT, Zhang Y, Su H, Jin Z, Xu SH, et al. Large scale free-standing open-ended TiO_2 nanotube arrays: Stress-induced self-detachment and in situ pore opening. Journal of Materials Chemistry C. 2013;**1**: 7498-7506. DOI: 10.1039/c3tc31642d

[100] Zhao M, Li J, Li Y, Wang J, Zuo Y, Jiang J, et al. Gradient control of the adhesive force between Ti/

TiO_2 nanotubular arrays fabricated by anodization. Scientific Reports. 2014;**4**:1-7. DOI: 10.1038/srep07178

[101] Alves SA, Rossi AL, Ribeiro AR, Toptan F, Pinto AM, Celis JP, et al. Tribo-electrochemical behavior of bio-functionalized TiO_2 nanotubes in artificial saliva: understanding of degradation mechanisms. Wear. 2017;**384-385**:28-42. DOI: 10.1016/j.wear.2017.05.005

[102] Alves SA, Rossi AL, Ribeiro AR, Toptan F, Pinto AM, Celis JP, et al. Improved tribocorrosion performance of bio-functionalized TiO_2 nanotubes under two-cycle sliding actions in artificial saliva. Journal of the Mechanical Behavior of Biomedical Materials. 2018;**80**:143-154. DOI: 10.1016/j.jmbbm.2018.01.038

[103] Alves SA, Rossi AL, Ribeiro AR, Werckmann J, Celis JP, Rocha LA, et al. A first insight on the bio-functionalization mechanisms of TiO_2 nanotubes with calcium, phosphorous and zinc by reverse polarization anodization. Surface and Coatings Technology. 2017;**324**:153-166. DOI: 10.1016/j.surfcoat.2017.05.073

[104] Mohedano M, Arrabal R, Mingo B. PEO of pre-anodized Al-Si alloys: Corrosion properties and influence of sealings. Applied Surface Science. 2015;**346**:57-67. DOI: 10.1016/j.apsusc.2015.03.206

[105] Dehnavi V, Li B, Shoesmith DW, Yang X, Rohani S. Effect of duty cycle and applied current frequency on plasma electrolytic oxidation (PEO) coating growth behavior. Surface & Coatings Technology. 2013;**226**:100-107. DOI: 10.1016/j.surfcoat.2013.03.041

[106] Rapheal G, Kumar S, Scharnagl N, Blawert C. Effect of current density on the microstructure and corrosion properties of plasma electrolytic oxidation (PEO) coatings on AM50 Mg alloy produced in an electrolyte containing clay additives. Surface & Coatings Technology. 2016;**289**:150-164. DOI: 10.1016/j.surfcoat.2016.01.033

[107] Matykina E, Arrabal R, Pardo A, Mohedano M, Mingo B. Energy-efficient PEO process of aluminium alloys. Materials Letter. 2014;**127**:13-16. DOI: 10.1016/j.matlet.2014.04.077

[108] Hussein RO, Nie X, Northwood DO. Production of high quality coatings on light alloys using plasma electrolytic oxidation (PEO). WIT Transactions on Built Environment. 2017;**166**:439-454. DOI: 10.2495/HPSM160411

[109] López-Ortega A, Arana JL, Rodríguez E, Bayón R. Corrosion, wear and tribocorrosion performance of a thermally sprayed aluminum coating modified by plasma electrolytic oxidation technique for offshore submerged components protection. Corrosion Science. 2018;**143**:258-280. DOI: 10.1016/j.corsci.2018.08.001

[110] Dehnavi V, Shoes mith DW, Luan BL, Yari M, Liu XY, Rohani S. Corrosion properties of plasma electrolytic oxidation coatings on an aluminium alloy e The effect of the PEO process stage. Materials Chemistry and Physics. 2015;**161**:49-58. DOI: 10.1016/j.matchemphys.2015.04.058

[111] Fatkullin AR, Parfenov EV, Yerokhin A, Lazarev DM, Matthews A. Effect of positive and negative pulse voltages on surface properties and equivalent circuit of the plasma electrolytic oxidation process. Surface & Coatings Technology. 2015;**284**:427-437. DOI: 10.1016/j.surfcoat.2015.07.075

[112] Kuromoto KN, Simão RE, Soares GA. Titanium oxide films produced on commercially pure titanium by anodic oxidation with different voltages. Materials Characterization.

2007;**58**(2):114-121. DOI: 10.1016/j.matchar.2006.03.020

[113] Saénz de Viteri V, Bayón R, Igartua A, Barandika G, Esteban Moreno J, Pérez-Jorge Peremarch C, et al. Structure, tribocorrosion and biocide characterization of Ca, P and I containing TiO_2 coatings developed by plasma electrolytic oxidation. Applied Surface Science. 2016;**367**:1-10. DOI: 0.1016/j.apsusc.2016.01.145

[114] Shokouhfar M, Dehghanian C, Montazeri M, Baradaran A. Preparation of ceramic coating on Ti substrate by plasma electrolytic oxidation in different electrolytes and evaluation of its corrosion resistance: Part II. Applied Surface Science. 2012;**258**:2416-2423. DOI: 10.1016/j.apsusc.2011.10.064

[115] Goriainov V, Cook R, Latham JM, Dunlop DG, Oreffo ROC. Bone and metal: An orthopaedic perspective on osseointegration of metals. Acta Biomaterialia. 2014;**10**:4043-4057. DOI: 10.1016/j.actbio.2014.06.004

[116] Alves SA, Bayón R, Igartua A, Saénz de Viteri V, Rocha LA. Tribocorrosion behavior of anodic titanium oxide films produced by plasma electrolytic oxidation for dental implants. Lubrication Science. 2013;**26**:500-513. DOI: 10.1002/ls.1234

[117] Alves SA, Bayón R, Saénz de Viteri V, García MP, Igartua A, Fernandes MH, et al. Tribocorrosion behavior of calcium and phosphorous-enriched titanium oxide films and study of osteoblast interactions for dental applications. Journal do Bio- and Tribo-Corrosion. 2015;**1**:1-23. DOI: 10.1007/s40735-015-0023-y

[118] Marques IDV, Barão VAR, da Cruz NC, Yuan JCC, Mesquita MF, Ricomini-Filho AP, et al. Electrochemical behavior of bioactive coatings on cp-Ti surface for dental application. Corrosion Science. 2015;**100**:133-146. DOI: 10.1016/j.corsci.2015.07.019

[119] Arena MA, Pérez-Jorge C, Conde A, Matykina E, Hernández-López JM, Pérez-Tanoira R, et al. Doped TiO_2 anodic layers of enhanced antibacterial properties. Colloids and Surfaces, B: Biointerfaces. 2013;**105**:106-112. DOI: 10.1016/j.colsurfb.2012.12.051

[120] Zhang E, Li F, Wang H, Liu J, Wang C, Li M, et al. A new antibacterialtitanium-copper sintered alloy: Preparation and antibacterial property. Materials Science and Engineering: C. 2013;**33**:4280-4287. DOI: 10.1016/j.msec.2013.06.016

[121] Yetim AF. Investigation of wear behavior of titanium oxide films, produced by anodic oxidation, on commercially pure titanium in vacuum conditions. Surface and Coatings Technology. 2010;**205**:1757-1763. DOI: 10.1016/j.surfcoat.2010.08.079

[122] Keronite. What is plasma electrolytic oxidation? [Internet]. 2017. Available from: https://blog.keronite.com/what-is-plasma-electrolytic-oxidation-article [Accessed: Oct 31, 2018]

[123] Ceschini L, Lanzoni E, Martini C, Prandstraller D, Sambogna G. Comparison of dry sliding friction and wear of Ti6Al4V alloy treated by plasma electrolytic oxidation and PVD coating. Wear. 2008;**264**(1-2):86-95. DOI: 10.1016/j.wear.2007.01.045

[124] Ravanetti F, Borghetti P, De Angelis E, Chiesa R, Martini FM, Gabbi C, et al. *In vitro* cellular response and *in vivo* primary osteointegration of electrochemically modified titanium. Acta Biomaterialia. 2010;**6**:1014-1024. DOI: 10.1016/j.actbio.2009.09.022

[125] Alves AC, Oliveira F, Wenger F, Ponthiaux P, Celis JP, Rocha LA. Tribocorrosion behaviour of

anodic treated titanium surfaces intended for dental implants. Journal of Physics D: Applied Physics. 2013;**46**(40):1-9. DOI: 10.1088/0022-3727/46/40/404001

[126] Ribeiro AR, Oliveira F, Boldrini LC, Leite PE, Falagan-Lotsch P, Linhares ABR, et al. Micro-arc oxidation as a tool to develop multifunctional calcium-rich surfaces for dental implant applications. Materials Science and Engineering: C. 2015;**54**:196-206. DOI: 10.1016/j.msec.2015.05.012

[127] Shirai T, Shimizu T, Ohtani K, Zen Y, Takaya M, Tsuchiya H. Antibacterialiodine-supported titanium implants. Acta Biomaterialia. 2011;**7**:1928-1933. DOI: 10.1016/j.actbio.2010.11.036

[128] Marques IDV, Alfaro MF, da Cruz NCZ, Mesquita MF, Takoudis C, Sukotjo C, et al. Tribocorrosion behaviour of biofunctional titanium oxide films produced by micro-arc oxidation: Synergism and mechanisms. Journal of the Mechanical Behavior of Biomedical Materials. 2018;**60**:8-21. DOI: 10.1016/j.jmbbm.2015.12.030

[129] Chen L, Wei K, Qu Y, Li T, Chang B, Liao B, et al. Characterization of plasma electrolytic oxidation film on biomedical high niobium-containing β-titanium alloy. Surface and Coatings Technology. 2018;**352**:295-301. DOI: 10.1016/j.surfcoat.2018.08.025

[130] Niinomi M, Nakai M, Hieda J. Development of new metallic alloys for biomedical applications. Acta Biomaterialia. 2012;**8**(11):3888-3903. DOI: 10.1016/j.actbio.2012.06.037

[131] Mattox DM. Handbook of Physical Vapor Deposition (PVD) Processing. 2nd ed. Norwich, NY, Oxford: William Andrew, Elsevier Science; 2010. ISBN: 978-0-8155-2037-5

[132] Martin PM. Handbook of Deposition Technologies for Films and Coatings. Science, Applications and Technology. 3rd ed. Elsevier Publishing; 2009. ISBN: 978-0-8155-2031-3

[133] Sanders DM, Anders A. Review od cathodic arc deposition technology at the start of the new millenium. Surface and Coatings Technology. 2000;**133—134**:78-90. DOI: 0.1016/S0257-8972(00)00879-3

[134] Sanchette F, Ducros C, Schmitt T, Steyer P, Billard A. Nanostructured hard coatings deposited by cathodic arc deposition: From concepts to applications. Surface and Coatings Technology. 2011;**205**:5444-5453. DOI: 10.1016/j.surfcoat.2011.06.015

[135] Wasa K, Kanno I, Kotera H. Handbook of Sputter Deposition Technology. 2nd ed. United States of America: Elsevier Publishing; 2012. ISBN: 978-1-4377-3483-6

[136] Brauer G, Szyszka B, Vergohl M, Bandorf R. Magnetron Sputtering-Milestones of 30 years. Vacuum. 2010;**84**:1354-1359. DOI: 10.1016/j.vacuum.2009.12.014

[137] Kelly PJ, Arnell RD. Magnetron sputtering: A review of recent developments and applications. Vacuum. 2000;**56**:159-172. DOI: 10.1016/S0042-207X(99)00189-X

[138] Sarakinos K, Alami J, Konstantinidis S. High power pulsed magnetron sputtering: A review on scientific and enginnering state of the art. Surface and coatings technology. 2010;**204**:1661-1684. DOI: 10.1016/j.surfcoat.2009.11.013

[139] Kouznetsov V, Macak K, Schneider JM, Helmersson U, Petrov I. A novel pulsed magnetron sputter technique utilizing very high target power densities. Surface and coatings

technology. 1999;**122**:290-293. DOI: 10.1016/S0257-8972(99)00292-3

[140] Helmersson U, Lattemann M, Bohlmark J, Ehiasarian AP, Gudmundsson JT. Ionized physical vapor deposition (IPVD): A review of technology and applications. Thin Solid films. 2006;**513**:1-24. DOI: 10.1016/j.tsf.2006.03.033

[141] Samuelsson M, Lundin D, Jensen J, Raadu MA, Gudmunsson JT, Helmersson U. On the film density using high power impulse magnetron sputtering. Surface and coatings technology. 2010;**205**(2):591-596. DOI: 10.1016/j.surfcoat.2010.07.041

[142] Antunes RA, de Oliveira MC. Corrosion processes of physical vapor deposition-coated implants. Critical Reviews in Biomedical Engineering. 2009;**37**(6):425-460. DOI: 10.1615/CritRevBiomedEng.v37.i6.10

[143] Ching HA, Choudhury D, Nine MJ, Osman NAA. Effects of surface coating on reducing friction and wear of orthopaedic implants. Science and technology of advanced materials. 2014;**15**:014402. DOI: 10.1088/1468-6996/15/1/014402

[144] Love CA, Cook RB, Harvey TJ, Deranley PA, Wood RJK. Diamond like carbon coatings for potential application in biological implants-a review. Tribology international. 2013;**63**:141-150. DOI: 10.1016/j.triboint.2012.09.006

[145] Donnet C, Erdemir A. Tribology of Diamond-like Carbon Films: Fundamentals and Applications. United State of America: Springer; 2008. ISBN: 978-0-387-30264-5

[146] Choudhury D, Lackner J, Fleming RA, Goss J, Chen J, Zou M. Diamond-like carbon coatings with zirconium-containing interlayers for orthopedic implants. Journal of the mechanical behavior of biomedical materials. 2017;**68**:51-61. DOI: 10.1016/j.jmbbm.2017.01.023

[147] Hatem A, Lin J, Wei R, Torres RD, Laurindo C, Biscaia de Souza G, et al. Tribocorrosion behavior of low friction TiSiCN nanocomposite coatings deposited on titanium alloy for biomedical applications. Surface & Coatings Technology. 2018;**347**:1-12. DOI: 10.1016/j.surfcoat.2018.04.049

[148] Sáenz de Viteri V, Barandika MG, Ruiz de Gopegui U, Bayón R, Zubizarreta C, Fernández X, et al. Characterization of Ti-C-N coatings deposited on Ti6Al4V for biomedical applications. Journal of Inorganic Biochemistry. 2012;**117**:359-366. DOI: 10.1016/j.jinorgbio.2012.09.012

[149] Sáenz de Viteri V, Barandika MG, Bayón R, Fernández X, Ciarsolo I, Igartua A, et al. Development of Ti-C-N coatings with improved tribological behavior and antibacterial properties. Journal of the Mechanical Behavior of Biomedical Materials. 2015;**55**:75-86. DOI: 10.1016/j.jmbbm.2015.10.020

[150] Bayón R, Igartua A, González JJ, Ruiz de Gopegui U. Influence of the carbon content on the corrosion and tribocorrosion perfromance of Ti-DLC coatings for biomedical alloys. Tribology International. 2015;**88**:115-125. DOI: 10.1016/j.triboint.2015.03.007

[151] Braic V, Braic M, Balaceanu M, Vladescu M, Zoita CN, Titorencu I, et al. (Zr,Ti)CN coatings as potential candidates for biomedical applications. Surface & Coatings Technology. 2011;**206**:604-609. DOI: 10.1016/j.surfcoat. 2011.03.074

[152] Schutze M, Bender R, Schutze KG. Corrosion Resistance of High-performance Materials. 2nd ed. Frankfurt: Wiley-VCH; 2012. ISBN: 978-3-527-33435-3

[153] Mendizabal L, López A, Bayón R, Herrero-Fernández P, Barriga J, González JJ. Tribocorrosion response in biological environments of multilayer TaN films deposited by HPPMS. Surface & Coatings Technology. 2016;**295**:60-69. DOI: 10.1016/j.surfcoat.2015.11.012

[154] Valero Vidal C, Igual Muñoz A. Electrochemical characterisation of biomedical alloys for surgical implants in simulated body fluids. Corrosion Science. 2008;**50**:1954-1961. DOI: 10.1016/j.corsci.2008.04.002

[155] H-Li H, Chang Y-Y, M-Cheng L, C-Rong L, C-Ho L, T-Ming S. Antibacterial TaN-Ag coatings on titanium dental implants. Surface & Coatings Technology. 2010;**205**:1636-1641. DOI: 10.1016/j.surfcoat.2010.07.096

[156] Narayanan R, Seshadri SK, Kwon TY, Kim KH. Calcium phosphate-based coatings on titanium and its alloys. Journal of Biomedical Materials Research Part B: Applied Biomaterials. 2008;**85**:279-299. DOI: 10.1002/jbm.b.30932

[157] Surmenev RA, Surmeneva MA, Ivanova AA. Significance of calcium phosphate coatings for the enhancement of new bone osteogenesis: A review. Acta Biomaterialia. 2014;**10**(2):557-579. DOI: 10.1016/j.actbio.2013.10.036

[158] Vranceanu DM, Cotrut CM, Bramowicz M, Titorencu I, Kulesza S, Kiss A, et al. Osseointegration of sputtered SiC-added hydroxyapatite for orthopaedic applications. Ceramic International. 2016;**42**:10085-10093. DOI: 10.1016/j.ceramint.2016.03.114

[159] Vladescu A, Birlik I, Braic V, Toparli M, Celik E, Ak Azem F. Enhancement of the mechanical properties of hydroxyapatite by SiC addition. Journal of the Mechanical Behavior of Biomedical Materials. 2014;**40**:362-368. DOI: 10.1016/j.jmbbm.2014.08.025

[160] Ak Azem F, Kiss A, Birlik I, Braic V, Luculescu C, Vladescu A. The corrosion and bioactivity behavior of SiC doped hydroxyapatite for dental applications. Ceramics International. 2014;**40**(10 A):15881-15887

Chapter 4

Bioprinting

Patrick Rider, Željka Perić Kačarević, Sujith Retnasingh, Said Alkildani, Ole Jung and Mike Barbeck

Abstract

Bioprinting is an emerging field in the areas of tissue engineering and regenerative medicine. It is defined as the printing of structures consisting of living cells, biomaterials and active biomolecules. The ultimate aim is to produce implantable organs and tissues to replace the use of autografts, which cause donor site morbidity and require two invasive surgeries. Not only is bioprinting aimed at the restoration of tissue, it has significant potential for drug delivery and cancer studies. Bioprinting provides control over cell placement and therefore creates a homogenous distribution of cells correlating to a uniform tissue ingrowth. Another attribute of bioprinting is the production of patient-specific spatial geometry, controllable microstructures and a high degree of reproducibility and scalability between designs. This book chapter will discuss the many parameters of bioprinting; manufacturing techniques, precursor materials, types of printed cells and the current research.

Keywords: inkjet printing, extrusion, stereolithography, laser-assisted, tissue engineering, hydrogel, bioink, additive manufacture, 3D printing

1. Introduction

There are more organ donors and recipients than there have ever been, however, in the UK, up to one in six patients requiring either a liver, heart or lung transplantation becomes too ill or dies while waiting for a donor organ [1]. When a donor organ becomes available, the recipient and clinicians have to make a fast decision whether to accept the organ, which can lead to an improper assessment of the associated risks. Risks associated with the donor organ include the age of the donor, retrieval of the organ after circulatory death and the potential to transmit blood-borne disease or cancer [1].

Bioprinting has the potential to eradicate the problems associated with organ donation and provide implantable organs on-demand. Bioprinting is a subcategory of additive manufacture, a process by which small scale objects are printed from a bottom up approach, through the deposition of successive layers of material. Using this approach enables the production of very precise and accurate designs and shapes to be built using minimal amounts of material. Designs are made on a computer which is then sent to a printer, meaning that designs are highly repeatable. The processes for developing a bioprinted tissue are outlined in **Figure 1**.

Although additive manufacturing uses a variety of materials to fabricate objects with different functions and purposes, bioprinting is specifically the printing of structures consisting of living cells, biomaterials and active biomolecules. First

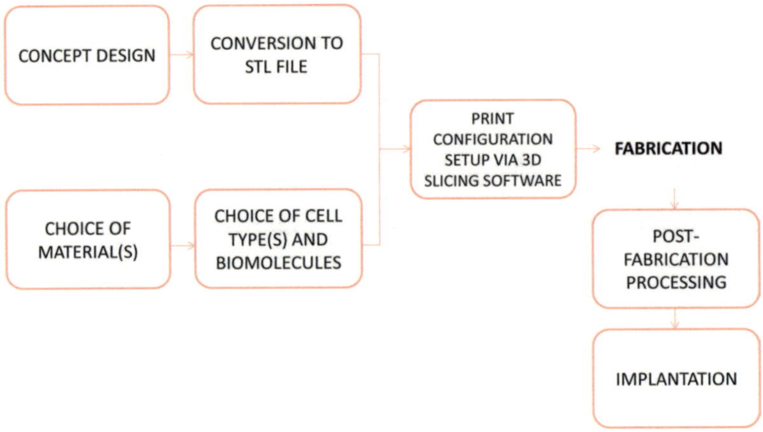

Figure 1.
Flow chart depicting the individual process involved with producing a bioprinted tissue.

patented attempts of bioprinting started with filling cartridges of table-top inkjet printers with bioinks, consisting mainly of cell-laden hydrogels [2]. Although research using inkjet technology was an early adapter for bioprinting, many other additive manufacturing techniques have also been utilized.

Common techniques include inkjet printing, stereolithography and extrusion printing, demonstrated in **Figure 2**. Although these techniques are well established, it is only in the last decade that these processes have been properly developed for applications in the field of regenerative medicine and tissue engineering. Compared to the conventional techniques of producing tissue engineering scaffolds, additive manufacture has the ability to consistently produce highly repeatable designs with a precise, well-defined micro and nanoscale structure [3, 4].

The ultimate goal of bioprinting is to produce an entire complex organ capable of being implanted. This will provide an alternative source of organs so that patients will no longer require long waiting periods to receive a donor organ. Additionally, by being able to fabricate organs on demand, the ethical issues surrounding the supply and use of human or animal tissues is removed.

Another possibility for bioprinting is to provide an alternative to animal testing. The potency and dosages of pharmaceuticals can be tested on bioprinted organoids or on an "organ-on-a-chip". This could prove to be a more efficient and effective

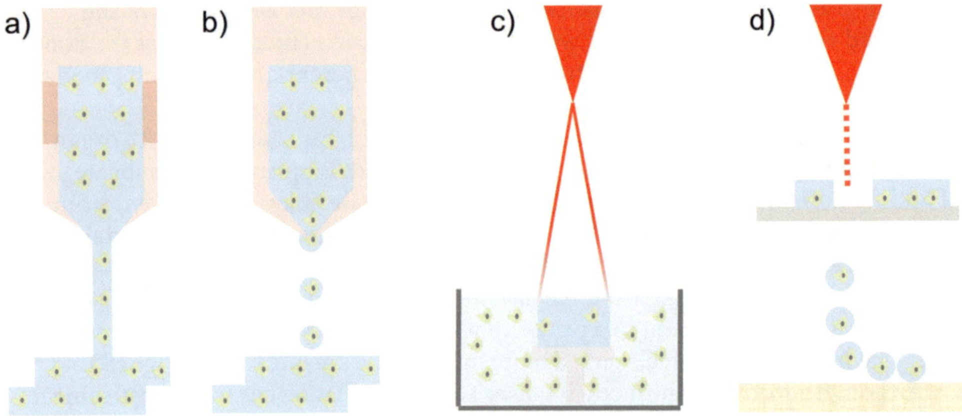

Figure 2.
Bioprinting techniques: (a) extrusion-based, (b) inkjet-based, (c) stereolithography and (d) laser-assisted.

method for the testing of pharmaceuticals, as the translation from an animal model in to humans is not without its flaws, as an animal system cannot fully replicate a human system.

Common tissue engineering manufacturing techniques are capable of producing tissue specific scaffolds; however they require to be seeded with cells post-fabrication. Cell seeding post-fabrication can lead to a poor distribution of cells, especially in larger scale 3D scaffolds. Homogeneous distribution of cells is not a problem with bioprinting, as printing can be used to control the positioning of cells during fabrication [5].

Simple tissues have been repaired in a clinical setting using tissue engineering scaffolds, such as the larynx [6], bladder [7] and urinary tract [8], however through bioprinting, the ability to build a more complex tissue scaffold for implantation becomes a greater prospect.

2. Printing techniques

2.1 Inkjet printing

Inkjet printing is a non-contact method of fabrication, which limits the risk of contamination during printing. Scaffolds are constructed using the deposition of droplets and can achieve excellent resolution ranging between 20 and 100 μm [9–11]. Droplets are ejected from a printhead via thermal, piezoelectric or microvalve processes. Thermal inkjet printing uses the nucleation of an air bubble to induce droplet formation, but a localized temperature between 100 and 300°C could inflict damage on cells and limits its use for printing natural polymers [10, 12]. Piezoelectric-based inkjet printing uses acoustic waves to eject a droplet, however this limits the use of viscous and therefore more concentrated inks, as these will dampen out the acoustic waves before they can expel a droplet [13]. Microvalve printing uses pneumatically controlled pistons to eject droplets. Inkjet printing can print multiple inks at once that can be used to interact with each other to control scaffold properties and create complex structures with multiple types of cells [14–16].

2.2 Extrusion

Bioink is extruded from a micro-nozzle to build a three-dimensional structure in layer-by-layer fashion. Extrusion-based printing is either mechanically (piston or screw) or pneumatically driven [9]. This fabrication technique uses highly viscous bioink and does not require any chemical additives to enable the curing of the material [9]. As inks are viscous, and therefore more concentrated, fabrication rates are relatively quick. However, the more viscous a bioink is, the higher the induced shear-stress during extrusion, which can result in a higher cell apoptotic activity. Out of the different extrusion bioprinting techniques, pneumatic systems have demonstrated the best cell viability post printing [17]. Extrusion-based techniques are promising as they create scaffolds with high structural integrity using bioinks containing high cell densities and can achieve resolutions around 200 μm [9, 18].

2.3 Laser-based printing

There are two laser-based techniques used for bioprinting. The first is stereolithography (SLA), which is a nozzle-free technique that depends on photocuring with ultraviolet light (UV). The UV energy is directed at a reservoir of photosensitive polymers to formulate 3D structures in a layer-by-layer fashion. This technique is known to have quick fabrication times as its speed depends on the height of the print rather than on the print complexity [19]. A resolution ranging between 5 and 300 μm can be achieved using SLA [20, 21]. This technique is not limited by the viscosity of the bioink, however, photoinitiators are often added to the bioink to improve polymer photosensitivity, and can affect the viability of the cells. The type and concentration of photoinitiator will influence cell viabilities differently [19]. Another concern is UV-induced damage to the cells, and has led to the development of alternative visible light-based and initiator-free techniques [22, 23]. Collectively, the short fabrication time and the absence of shear stress-induced apoptosis make SLA a good candidate for bioprinting.

Another laser-based bioprinting technique is laser-assisted bioprinting (LAB), which is a scaffold-free and nozzle-free technique where droplets of biomaterial/cells/peptides are propelled from a donor material slide onto a receiver material. Laser energy induces a vaporization effect used to transfer material from the donor material slide onto the receiver. This technique can achieve a micro-scale resolution with multicell positioning [24, 25]. The accurate positioning of multiple cell types using this technique has great potential in creating biologically relevant complex designs, yet the main drawbacks of this technique are its low stability and scalability [24].

3. Bioinks

Bioink is a term given to the precursor material used during printing. The main component of the bioink usually consists of a hydrogel. Hydrogels are used as they are highly hydrated, enable cell encapsulation and provide a cushioning effect during fabrication. Furthermore, hydrogels have adaptable rheological properties making them suitable for processing and printing [15]. Hydrogels used in tissue engineering applications are either naturally derived or synthetically derived polymers.

3.1 Naturally derived hydrogels

Collagen, gelatin (denatured collagen), silk, alginate and chitosan are the most commonly used naturally derived polymers in bioprinting with collagen being used in 26% of the bioprinting-relevant literature [26]. Natural polymers provide great bioactivity and high adhesiveness, resulting in high cell viability and proliferation. However, a lack of mechanical competence and reproducibility limit their use in a non-composite form.

3.2 Synthetically derived hydrogels

Aliphatic polyesters and poly(ethylene glycol) are the most commonly used synthetic polymers in bioprinting [27]. Synthetic polymers have a high printing fidelity and controllable degradation kinetics and mechanical properties. However, their inert properties result in low cell viability and proliferation rate [28]. Aliphatic polyesters used in tissue engineering are usually poly(lactic acid), poly(glycolic acid) and poly(caprolactone). Refer to **Table 1** for a comparison of their different properties.

3.3 Characteristics

When processing bioinks, three main parameters must be considered: viscosity of the precursor hydrogel, gelation and network stiffness post-processing [30].

Aliphatic polyester	Hydrophilicity/hydrophobicity	Degradation time	Young's modulus [29]
PCL	Very hydrophobic	Lowest	0.4 GPa
PLA	Hydrophobic	Medium	2.7 GPa
PGA	Hydrophilic	Fast	7.0 GPa
PLGA	Ratio of different polymers will influence rate of degradation and water affinity; PLA percentage means mixture has lower degradation rate and is more hydrophobic.		Amorphous

Table 1.
Correlation between water affinity and degradation rate for aliphatic polyesters.

The latter two points can affect cell viability, proliferation, migration and even differentiation. The three parameters are mainly affected by the type of polymer, its concentration and molecular weight. Other factors that can have an adverse effect on cells are the shear stress induced during printing and the time period during which the cells are exposed to non-physiological conditions (fabrication time).

Rheology of the bioink, particularly viscosity, is an important consideration for printing. Viscosity affects printing fidelity and also influences the shear stresses induced during printing. High shear stresses during printing can damage cells in nozzle-based techniques leading to higher cell apoptotic activity. Each printing technique requires a different viscosity range; inkjet-based printers require low viscosity solutions [31], extrusion-based printers require a high viscosity [9], while laser-assisted bioprinting requires a medium viscosity [5]. However, stereolithographic techniques are not as limited by hydrogel viscosity ranges. Recent research has shown that the sheer stresses induced by the applied pressure in extrusion printing can be alleviated if the bioinks are prepared as isotonic solutions [32].

The objective to have a 3D structure laden with cells that are viable, proliferative and differentiative, along with a structure that exhibits mechanical competence, stability and biologically relevant complexity is dependent on the bioink and the printing process. Ultimately, the type of polymer, its chemical properties, viscosity, gelation, stiffness and fabrication time will affect the cells' status and the print fidelity.

4. Bioprinted tissues

4.1 Neural

Injuries of the central and peripheral nervous system can be challenging to repair. The central nervous system does not regenerated under normal conditions, and any injuries it sustains can lead on to neurogenerative disease. Damage to the peripheral nervous system is mostly treated with autografts and allografts, however motor and sensory recovery rates are poor. In a review of studies recording sensory recovery after digital nerve repair, only 67% of patients reported good sensory repair or better with a nerve graft, which increased to 79% of patients with artificial conduits [33]. This result indicates the potential for synthetically made nerve guidance conduits and therefore the use of bioprinting.

Some of the initial work involving bioprinting of mammalian cells was performed using Chinese hamster ovary (CHO) and rat embryonic motoneuron cells [34]. It was established that using an inkjet printer to print the cells did not affect cell proliferation or the polarized morphology of the motoneuron cells. However, evaluation of cell lysis immediately after printing indicated that ~8% of the cells were damaged due to the printing process. Further work with rat primary embryonic hippocampal and cortical neurons demonstrated a cell viability of 74.2 ± 6.3%, as well as the maintenance of basic function, phenotype and electrophysiological properties [35].

Lorber et al. inkjet printed cells from the mature adult nervous system [36]. Monitoring the cells during droplet ejection, the cells did not undergo major distortion and hence destruction caused by shear forces. It was shown that printed glia cells retained their ability to support the growth of the seeded retinal ganglion cells, which is an essential interaction for maintaining proper nerve function. Tse et al. printed a combination of neuronal and glial cells [37]. High cell viabilities >86% for neuronal cells and >89% for Schwann cells were achieved. Not only did the printed cells exhibit early neurite growth in comparison to non-printed cells, neurites also grew longer in length.

Using an extrusion printing process, neural stem cells (NSCs) have been printed embedded in polyurethane (PU) [38]. The bioprinted scaffolds restored nerve function when used to treat Zebra fish with a traumatic brain injury. Owens et al. developed the extrusion bioprinting process to produce purely cellular nerve grafts [39, 40]. Mouse bone marrow stem cells (BMSCs) and Schwann cells (SC) were extruded alongside agarose rods. The agarose rods were removed after a 7 day maturation period leaving a tubular cell structure. Implanted to repair a sciatic nerve injury in rats, both motor and sensory function was restored after 40 weeks.

SLA has been used to print human neural stem cells (hNSCs) [41]. Scaffolds were made containing GelMA and graphene nanoplatelets. GelMA content had a negative effect on cell proliferation rates, while the presence of graphene, known to stimulate hNSCs into neural differentiation [42], improved the cellular response. By monitoring glial fibrillary acidic protein (GFAP) expression, which is associated with the development of ependymal cells, and ß-tubulin III, which is associated with neurons, it was determined that the printed cells had differentiated over a 14 day period post-printing.

4.2 Skin

In the repair of deep partial thickness and full thickness burns a new epidermal barrier needs to be established, this is usually achieved through transplantation [43]. Current treatment techniques, including split thickness grafts, lack the inclusion of structures such as hair follicles, subcutaneous glands or sweat glands and do not provide the full restoration of sensory and motor neurons [43]. This can cause problems for the patient, especially the lack of sweat glands, which can prevent proper thermo-regulation. Bioprinting can be used to develop a complex tissue as multiple inks can be printed at once. Although the production of replica human skin has yet to be achieved, there have been many promising advances for the treatment of full thickness wounds.

Laser-assisted bioprinting was tested for the printing of different skin cells: fibroblasts and keratinocytes, as well as human mesenchymal stem cells (hMSCs) [25]. Printing was shown not to induce apoptotic activity and cell viabilities were measured as >90%. The hMSCs maintained their phenotype, implying that the transfer process did not induce differentiation of the cells.

Kim et al. bioprinted a skin model using both extrusion and inkjet technologies [44]. A collagen/PCL ink was extruded to form the basis of the scaffold, followed by the inkjet printing of primary human dermal fibroblasts (hFBs) and primary human epidermal keratinocytes (hKCs). Using inkjet printing to deposit the cells provided a controllable and even distribution. When placed in cell culture over a

14 day period, the bioprinted construct developed biological characteristics such as elongated hFB, and dermal and epidermal layers.

A PEG based skin model was produced by Rimann et al. [45] using a microvalve droplet system. Alternate layers were printed of a PEG-based ink followed by the hFBs ink suspension. Each layer of the PEG based ink was immediately polymerized using UV light. Due to the use of UV light the cells were analyzed for UV damage. By comparing printed and unprinted cells, it was shown that the cell DNA was not damaged during the printing process. The cells adhered to the structure, maintaining a typical morphology and were well dispersed within the polymer matrix. After a 3-week period, the cells continued to proliferate and produce their own extra cellular matrix, however, subsequent seeding of hKCs on top of the printed scaffolds did not achieve a fully stratified epidermis when exposed to the air-liquid interface for 14 days.

A more complex bioprinted skin model has been achieved by printing multiple cell types [16, 46, 47]. Using a microvalve droplet system, initial investigatory work used both hFBs and hKCs suspended in cell media that were printed in-between layers of a printed collagen gel [46, 47]. Using a droplet based technique enabled the precise placement of cells, with a recorded 68 ± 13 hKCs per droplet and 93 ± 13 hFBs per droplet for each respective ink suspension. Compared to skin tissue scaffolds that had used a manual seeding technique, the bioprinted scaffolds retained their shape better and developed denser epidermal layers. The epidermal layers formed tight junctions between neighboring cells, an important feature of the epidermis for providing an effective barrier.

Further work included pigmentation of the printed skin model by printing human epidermal melanocytes (hMCs) into the epidermal layer [16], demonstrated in **Figure 3**. The hKCs reached a terminal differentiation to form a stratum corneum-like structure. hMCs started producing melanin which accumulated at hKC interfaces, causing a light pigmentation. When the hMCs were printed as a uniform layer, they formed nevus-like structures (freckles), however it was suggested that this could be prevented by controlling hMC cell densities to create a more even spatial distribution.

A collagen skin scaffold containing hFCs and printed human dermal microvascular endothelial cells (hMVECs), that was subsequently seeded with neonatal hKCs mixed in collagen, was compared to a commercial graft (Apligraf®) for treating full thickness skin wounds in mice [48]. After 14 days, the bioprinted scaffolds had adhered to the wound while the Apligraf® had dried out and detached from the implant site, although no infections were recorded for any of the test groups. Complete closure of the wound occurred first for the bioprinted scaffold after 21 days, followed by the Apligraf® after 28 days. Wound sites treated with either the commercial graft or the bioprinted graft developed epidermal and dermal

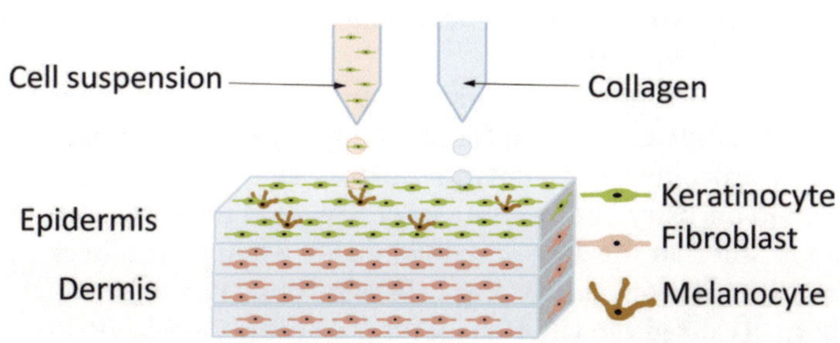

Figure 3.
Bioprinted pigmented skin model fabricated using a microvalve droplet technique [16]. Fibroblasts are used for a dermal layer, and keratinocytes and melanocytes for an epidermal layer.

layers, however the bioprinted skin graft was shown to have the most histologically similar appearance to native skin, and developed a micro-vessel network.

Skardal et al. successfully demonstrated *in-situ* bioprinting to repair full thickness skin wounds on the backs of mice [49]. Amniotic fluid-derived stem (AFS) cells and bone marrow-derived mesenchymal stem cells (MSCs) were separately suspended in fibrinogen-collagen solutions and printed into the wound sites. The fibrinogen-collagen solution was crosslinked by printing a thrombin solution between each layer. Printing the skin graft directly into the wound site achieved complete coverage of the wound and a tight seal between the skin and graft. At all points during the study, wound contraction and re-epithelialization were significantly higher in wound sites treated with bioprinted cells. Printed scaffolds exhibited re-epithelialization of up to 89% which had organized, well defined layers, while the control only exhibited a 51% re-epithelialization with a poor structural quality.

4.3 Cardiovascular

Cardiovascular disease is one of the major causes of death globally. According to World Health Organization (WHO), cardiovascular disease accounts for more than one third of the total deaths each year [50, 51]. Depending on severity, treatment solutions may require replacement of damaged vital vasculature, or even heart transplantation. Problems such as organ shortages or requirements for multiple invasive surgeries mean that alternative solutions need to be found.

Different methods have been explored for the bioprinting of a 3D vascular network. Endothelial cells have been successfully printed with a high degree of accuracy using an inkjet printer. It was shown that during printing, a precise number of cells were ejected per drop [52]. Boland et al. used inkjet printing to print a crosslinking agent (calcium chloride) into an alginate/gelatin solution to produce vascular structures [53]. By printing a crosslinking agent into the polymer solution, a structure was formed out of hollow shells providing a microscopic porosity that proved beneficial for cell migration.

Cui et al. used inkjet printing to encapsulate human microvascular endothelial cells (hMVEC) in a fibrin hydrogel [54]. Scaffolds were fabricated with micron-sized fibrin channels that had similar mechanical properties to other tissue engineered blood vessels. In cell culture, the cells aligned along the channels in a confluent lining, and had formed a ring-shaped microvasculature that sealed the fibroin channel with a high level of integrity.

Extrusion bioprinting has been used to print heart valve conduits [55]. Scaffolds were fabricated using a bioink of methacrylayed hyaluronic acid (Me-HA) and methacrylated gelatin (Me-Gel) with suspended human aortic valvular interstitial cells (hAVIC). In cell culture, cells adhered and formed a monolayer on the surface of the bioprinted structure while the encapsulated cells below the surface started to remodel the hydrogel after 3 days.

Initial work into printing cardiac tissue was performed using alginate hydrogel and primary feline adult and human H1 cardiomyocytes [56]. Inkjet printed in a half heart shape, scaffolds had a 1 cm inner diameter and two connected ventricles. Due to the printed porosity, the large structure enabled the transportation of nutrients, and meant that the cells remained viable when placed in cell culture. Using electrical stimulation, the cardiac cells were observed to contract rhythmically which in turn caused the whole structure to beat periodically.

The generation and use of human pluripotent stem cell (iPSCs) is of growing interest for many applications as they can differentiate into multiple cell lineages [50]. Zhang et al. demonstrated a bioprinting methodology to fabricate endothelialized myocardium using iPSCs [57]. GelMA and iPSC scaffolds were extruded and seeded with cardiomyocytes to induce the formation of myocardium. The fabricated scaffolds were capable of spontaneous and synchronous contraction using low level electrical stimuli.

4.4 Bone

An aging population and the use of bone grafts to augment bone for repair and regeneration, have led to an increasing demand of functional bone grafts. The current gold standard of treatment involves autologous bone retrieved from the patients' iliac crest, however this requires multiple surgery sites, is expensive to perform and increases the associated risks involved with a second surgery site [58]. Therefore an alternative solution, such as the development of bioprinted bone grafts, presents an attractive alternative.

Using SLA, stem cells derived from mouse bone marrow stromal cells (OP-9 cells) were embedded in crosslinked PEGDA hydrogels [59]. After 24 h, the cells remained viable and subsequent seeding of mouse mesenchymal stem cells (mMSCs) onto the scaffold showed intensive mineralization.

Catros et al. used LAB to position osteosarcoma cells (MG63) onto poly(caprolactone) (PCL) scaffolds [60]. Cells were printed either between layers of electrospun PCL or only as a final layer. *In vitro* testing initially showed no difference in proliferation rate between the two groups, however, after 14 days, the scaffolds that included printed cells between layers of PCL exhibited a significantly higher proliferation rate. When implanted in calvarial defects in mice, significant bone tissue ingrowth was observed that had an organized structure. Extruded PCL scaffolds containing decellularized bone matrix and human adipose-derived stem cells (hASCs) used for craniofacial regeneration in mice, showed improved bone regeneration in comparison to a pure PCL scaffold [61].

Using a bioink consisting of either a matrigel or matrigel and alginate, endothelial progenitor cells (EPCs) and multipotent stromal cells (MSCs) were extruded into bone tissue scaffolds [62]. Implanted into subcutaneous dorsal pockets of mice, the scaffolds without the inclusion of alginate performed better, as they developed a bone-like tissue structure.

Zhou et al. employed a table-top stereolithographic printer to create biomimetic bone matrices to study metastasis of bone cancer in bone tissue [63], as shown in **Figure 4**. Scaffolds containing osteoblasts and hMSCs suspended in hydrogels containing different concentrations of GelMA and nanocrystalline hydroxyapatite, were subsequently seeded with breast cancer cells (BrCas). Placed in cell culture, osteoblast proliferation rates were observed to decrease, while BrCas proliferation rates increased. Interactions between the cell types within the bioprinted model led to the conclusion that the osteoblasts and MSCs secrete macromolecules that promote BrCas growth.

4.5 Cartilage

Because of its avascular nature and the presence of low cell densities, cartilage defects will not completely self-regenerate [64]. It is therefore important that different treatment possibilities are investigated to initiate its' regeneration.

Xu et al. fabricated a loadbearing cartilage scaffold using an inkjet printer to bioprint rabbit articular chondrocytes suspended in a fibrin–collagen hydrogel onto an electrospun PCL matrix [65]. It was found that by including a PCL fibrous network within the scaffold, mechanical properties were significantly increased. Scaffolds implanted subcutaneously in mice developed a dense, well organized

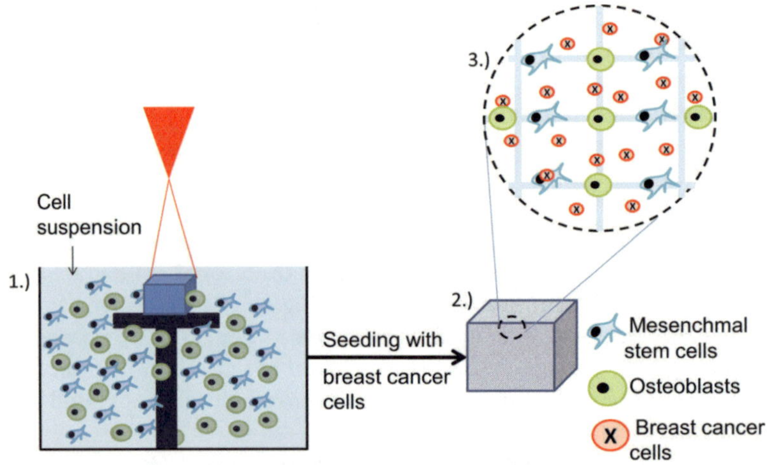

Figure 4.
Stereolithographic process to fabricate a cell-laden bone model to study breast cancer metastasis. (1) Homogenous suspension of MSCs and osteoblasts in a photocurable hydrogel; (2) cell-laden 3D structure that is seeded with breast cancer cells; (3) a close-up of homogeneously adhering MSCs/osteoblasts and a suspension of breast cancer cells.

collagen, not present in the control group (unseeded scaffolds). The cartilage tissue formation within the bioprinted scaffolds appeared histologically similar to normal elastic cartilage.

The use of PCL to provide structural support was also employed by Schuurman et al., who extruded bioprinted GelMA embedded with chondrocytes [66]. For improved cartilage development, scaffolds were also fabricated with a GelMA/hyaluronic acid hydrogel. The inclusion of hyaluronic acid improved cell viability *in vitro*, however when implanted in mice, both scaffolds types achieved similar levels of cartilage tissue formation. Other extruded PCL-alginate gel scaffolds that contained chondrocytes also reported the formation of cartilage after being implanted into the dorsal subcutaneous space of mice [67].

In situ bioprinting has been investigated by Cui et al. using human chondrocytes suspended in a poly(ethylene glycol) dimethacrylate (PEGDMA) solution [68, 69]. Scaffolds were printed directly into osteochondral plugs that had been harvested from bovine femoral condyles. Compared to scaffolds fabricated outside of the defect, the *in situ* bioprinted scaffolds had enhanced tissue integration. Another *in situ* bioprinting method has been developed by O'Connell et al. [70]. Based on an extrusion bioprinting technique, an "*in-situ* biopen" has been made comprising of two ink chambers and an extruder nozzle with a UV source that enables direct application of a bioink during surgery. Using the biopen, a human infrapatellar fat pad adipose stem cell (IPFP) laden gelatin-methacrylamide/hyaluronic acid-methacrylate (GelMA/HAMa) hydrogel scaffold was fabricated for *in vitro* testing. It was observed that the viability of biopen printed cells was 97% after 7 days.

4.6 Muscle

Bioprinting has shown great potential for building highly hierarchically organized cellular structures that comprise muscle tissue. Miri et al. created hierarchical cell-laden structures to mimic multicellular tissues [71]. Using extrusion-based bioprinting, multiple bioinks were deposited onto microfluidic chips. Structures resembling musculoskeletal junctions were printed using poly(ethylene glycol) diacrylate (PEGDA) and GelMA that were loaded with NIH/3T3 fibroblasts and C2C12 skeletal muscle cells. *In vitro*, the chips retained their printed interfaces and demonstrated adequate proliferation rates. PEGDA-framed chips that had a concentration-gradient of methacrylated gelatin (GelMA) ranging from 5 to 15%, were implanted subcutaneously in rats. After 30-days, enhanced cell proliferation was reported for regions containing 10% GelMA, signifying a great potential for angiogenesis. This study demonstrates the possibility to create hierarchically cell-laden structures to mimic multicellular tissues.

Bajaj et al. used stereolithography to bioprint muscle tissues composed of mouse embryonic stem cells (mESCs) and mouse myoblast cell line (C2C12), embedded in PEGDA [72]. Live/dead staining showed a slight decrease in cell viability for the bioprinted scaffolds, possibly due to the non-adhesive properties of PEGDA.

4.7 Periodontal

Periodontitis is a chronic inflammatory condition resulting in total destruction of the periodontium consists of alveolar bone, cementum, gingiva and periodontal ligaments and if left untreated, can result in tooth loss [73, 74]. Periodontal ligament stem cells (PDLSCs) have been shown to support the regeneration of periodontal tissues [75]. Ma et al. investigated cell viability of extruded PDLSCs encapsulated in a GelMA/PEG hydrogel [76]. It was shown that cell viabilities were around 94% after 24 hours post printing. Extrusion printing has also been used to reconstruct the maxillary bone in a 12-year-old dog, and is believed to be the first case of maxillary bone reconstruction using a 3D printed scaffold [77].

4.8 Corneal

Diseases of the cornea have a significant impact on visual health worldwide. Corneal opacity is the fourth leading cause of bilateral blindness globally [78, 79]. A total of approximately 50,000 corneal transplantations were performed in the United States in 2013 [80–82], yet drawbacks to this procedure include a reduced quality of visual recovery due to an early endothelial cell loss, detachment of the posterior lamellar grafts and vascular in-growth into the lamellar plane [79]. Using sodium alginate and methacrylated type I collagen mixed with corneal keratinocytes, Isaacson et al. extruded corneal scaffolds [83]. To produce a concave structure, gelatin was used as a support material and had a hollowed out shape. Cell viability of cornel keratinocytes were 90% after 1 day post-fabrication which dropped to 83% after 7 days.

The limbus borders the cornea and provides it with limbal epithelial stem cells (LESCs) for regeneration. Damage caused to it by disease or injury can impair a person's vision [84]. Using laser-assisted bioprinting Sorkio et al. printed human embryonic stem cell derived limbal epithelial stem cells (hESCs-LESCs) and hASCs for the repairing of the limbus [85]. *Ex vivo* assessment of the structures was performed on porcine corneal tissue, with results compared to commercially available acellular Matriderm® sheets as a control. The bioprinted scaffolds exhibited strong adhesion with the host tissue enabling hASCs migration, while control group had a more limited response.

5. Conclusion

Bioprinting has been used to create a variety of complex tissues and has demonstrated great potential as an alternative to autologous, allogeneic and xenogeneic organ and tissue transplantation. Progress towards a complete implantable organ is at various stages of development depending on the organ or tissue to be fabricated.

Some of the greatest developments have been with neural and skin tissues. It has been shown that neuronal cells can be printed without affecting their function or phenotype. In one investigation using an inkjet printer, the printed neuronal cells demonstrated faster and longer neurite growth in comparison to non-printed cells [37]. When implanted, bioprinted neural scaffolds have been able to restore nerve function [38–40].

Complex structures such as skin have been fabricated that include both dermal and an epidermal layers. By using bioprinting technology, it is possible to create a skin model with pigmentation [16]. Bioprinted skin scaffolds have been successfully applied via direct *in situ* printing [48]. Used to repair full thickness wounds in mice, *in situ* bioprinted scaffolds have regenerated skin tissue with a histologically similar appearance to native skin.

A potential use for bioprinted tissues could be in modeling diseased tissues. Research has already been conducted studying the metastasis of bone cancer in a bioprinted bone tissue model [63]. Using bioprinted tissue models it could be possible to gain a greater understanding of the interactions between infected and diseased cells with healthy tissues. By doing so, new treatments can be produced and tested to determine the most effective solution.

Each bioprinting technique has demonstrated the potential to generate fully functional organs, which in the future can be used for applications such as: direct implantation, to fill the gap for much needed donor organs; provide a model system for pharmaceutical testing, and replace the necessity for animal testing; or be used for disease modeling and enable different treatments to be easily explored. Overall, bioprinting is gaining substantial interest in the field of tissue engineering and regenerative medicine due to its capability to produce complex organs and tissues.

Conflict of interest

The authors of this book chapter have no conflicts of interest to declare.

Author details

Patrick Rider[1], Željka Perić Kačarević[2]*, Sujith Retnasingh[3,4], Said Alkildani[5], Ole Jung[6] and Mike Barbeck[6,1]

1 Botiss Biomaterials, Berlin, Germany

2 Department of Anatomy Histology, Embryology, Pathology Anatomy and Pathology Histology, Faculty of Dental Medicine and Health, University of Osijek, Osijek, Croatia

3 Institute for Environmental Toxicology, Martin-Luther-Universität, Halle-Wittenberg, Germany

4 Faculty of Biomedical Engineering, Anhalt University of Applied Science, Köthen, Germany

5 Department of Biomedical Engineering, Faculty of Applied Medical Sciences, German-Jordanian University, Amman, Jordan

6 Department of Oral and Maxillofacial Surgery, University Hospital Hamburg-Eppendorf, Hamburg, Germany

*Address all correspondence to: zeljkapericc@gmail.com

References

[1] Neuberger J, Trotter P, Stratton R. Organ transplantation rates in the UK. BMJ (Clinical Research Ed). 2017;**359**:j5218

[2] Wilson WC, Boland T. Cell and organ printing 1: Protein and cell printers. Anatomical Record-Part A Discoveries in Molecular, Cellular, and Evolutionary Biology. 2003;**272**:491-496

[3] Yeong WY, Chua CK, Leong KF, et al. Rapid prototyping in tissue engineering: Challenges and potential. Trends in Biotechnology. 2004;**22**:643-652

[4] Stevens B, Yang Y, Mohandas A, et al. A review of materials, fabrication methods, and strategies used to enhance bone regeneration in engineered bone tissues. Journal of Biomedical Materials Research—Part B Applied Biomaterials. 2008;**85**:573-582

[5] Murphy SV, Atala A. 3D bioprinting of tissues and organs. Nature Biotechnology. 2014;**32**:773-785

[6] Omori K, Tada Y, Suzuki T, et al. Clinical application of in situ tissue engineering using a scaffolding technique for reconstruction of the larynx and trachea. Annals of Otology, Rhinology & Laryngology. 2008;**117**:673-678

[7] Atala A, Bauer SB, Soker S, et al. Tissue-engineered autologous bladders for patients needing cystoplasty. The Lancet. 2006;**367**:1241-1246

[8] Raya-Rivera A, Esquiliano DR, Yoo JJ, et al. Tissue-engineered autologous urethras for patients who need reconstruction: An observational study. The Lancet. 2011;**377**:1175-1182

[9] Pati F, Jang J, Lee JW, et al. Extrusion bioprinting. 2015. DOI: 10.1016/B978-0-12-800972-7.00007-4

[10] Guo Y, Patanwala HS, Bognet B, et al. Inkjet and inkjet-based 3D printing: Connecting fluid properties and printing performance. 2017;**3**:562-576

[11] Liu F, Liu C, Chen Q, et al. Progress in organ 3D bioprinting. International Journal of Bioprinting. 2018;**4**:1-15

[12] Jammalamadaka U, Tappa K. Recent advances in biomaterials for 3D printing and tissue engineering. Journal of Functional Biomaterials. 2018;**9**:22

[13] Tekin E, Smith PJ, Schubert US. Inkjet printing as a deposition and patterning tool for polymers and inorganic particles. Soft Matter. 2008;**4**:703

[14] Smith PJ, Morrin A. Reactive inkjet printing. Journal of Materials Chemistry. 2012;**22**:10965

[15] Rider P, Zhang Y, Tse C, et al. Biocompatible silk fibroin scaffold prepared by reactive inkjet printing. Journal of Materials Science. 2016;**51**:8625-8630

[16] Zeng M, Jin S, Ye K. Tissue and organ 3D bioprinting. SLAS Technology: Translating Life Sciences Innovation. 2018;**23**(4):301-314. DOI: 10.1177/2472630318760515

[17] Ozbolat IT, Hospodiuk M. Current advances and future perspectives in extrusion-based bioprinting. Biomaterials. 2016;**76**:321-343

[18] Bishop ES, Mostafa S, Pakvasa M, et al. 3-D bioprinting technologies in tissue engineering and regenerative medicine: Current and future trends. Genes and Diseases. 2017;**4**:185-195

[19] Mondschein RJ, Kanitkar A, Williams CB, et al. Polymer

structure-property requirements for stereolithographic 3D printing of soft tissue engineering scaffolds. Biomaterials. 2017;**140**:170-188

[20] Raman R, Bhaduri B, Mir M, et al. High-resolution projection microstereolithography for patterning of neovasculature. Advanced Healthcare Materials. 2016;**5**:610-619

[21] Yue K, Trujillo-de Santiago G, Alvarez MM, et al. Synthesis, properties, and biomedical applications of gelatin methacryloyl (GelMA) hydrogels. Biomaterials. 2015;**73**:254-271

[22] Wang Z, Abdulla R, Parker B, et al. A simple and high-resolution stereolithography-based 3D bioprinting system using visible light crosslinkable bioinks. Biofabrication. 2015;7:45009

[23] Hoffmann A, Leonards H, Tobies N, et al. New stereolithographic resin providing functional surfaces for biocompatible three-dimensional printing. Journal of Tissue Engineering. 2017;**8**:204173141774448

[24] Guillotin B, Souquet A, Catros S, et al. Laser assisted bioprinting of engineered tissue with high cell density and microscale organization. Biomaterials. 2010;**31**:7250-7256

[25] Koch L, Kuhn S, Sorg H, et al. Laser printing of skin cells and human stem cells. Tissue Engineering. Part C, Methods. 2010;**16**:847-854

[26] Carrow JK, Kerativitayanan P, Jaiswal MK, et al. Polymers for Bioprinting. Elsevier Inc; 2015. DOI: 10.1016/B978-0-12-800972-7.00013-X

[27] Cameron DJA, Shaver MP. Aliphatic polyester polymer stars: Synthesis, properties and applications in biomedicine and nanotechnology. Chemical Society Reviews. 2011;**40**:1761-1776

[28] Delaittre G, Greiner AM, Pauloehrl T, et al. Chemical approaches to synthetic polymer surface biofunctionalization for targeted cell adhesion using small binding motifs. Soft Matter. 2012;**8**:7323-7347

[29] Gunatillake PA, Adhikari R, Gadegaard N. Biodegradable synthetic polymers for tissue engineering. European Cells and Materials. 2003;**5**:1-16

[30] Malda J, Visser J, Melchels FP, et al. 25th anniversary article: Engineering hydrogels for biofabrication. Advanced Materials. 2013;**25**:5011-5028

[31] Rider P, Brook I, Smith P, et al. Reactive inkjet printing of regenerated silk fibroin films for use as dental barrier membranes. Micromachines. 2018;**9**:46

[32] Godar DE. 3D bioprinting: Surviving under pressure. In: Tissue Regeneration. InTech; 2018. DOI: 10.5772/intechopen.73137

[33] Paprottka FJ, Wolf P, Harder Y, et al. Sensory recovery outcome after digital nerve repair in relation to different reconstructive techniques: Meta-analysis and systematic review. Plastic Surgery International. 2013;**2013**:704589

[34] Xu T, Jin J, Gregory C, et al. Inkjet printing of viable mammalian cells. Biomaterials. 2005;**26**:93-99

[35] Xu T, Gregory CA, Molnar P, et al. Viability and electrophysiology of neural cell structures generated by the inkjet printing method. Biomaterials. 2006;**27**:3580-3588

[36] Lorber B, Hsiao WK, Hutchings IM, et al. Adult rat retinal ganglion cells and glia can be printed by piezoelectric inkjet printing. Biofabrication. 2014;**6**. DOI: 10.1088/1758-5082/6/1/015001

[37] Tse C, Whiteley R, Yu T, et al. Inkjet printing Schwann cells and neuronal analogue NG108-15 cells. Biofabrication. 2016;**8**: 0

[38] Hsieh FY, Lin HH, Hsu SH. 3D bioprinting of neural stem cell-laden thermoresponsive biodegradable polyurethane hydrogel and potential in central nervous system repair. Biomaterials. 2015;**71**:48-57

[39] Owens CM, Marga F, Forgacs G, et al. Biofabrication and testing of a fully cellular nerve graft. Biofabrication. 2013;**5**. DOI: 10.1088/1758-5082/5/4/045007

[40] Marga F, Jakab K, Khatiwala C, et al. Toward engineering functional organ modules by additive manufacturing. Biofabrication. 2012;**4**. DOI: 10.1088/1758-5082/4/2/022001

[41] Zhu W, Harris BT, Zhang LG. Gelatin methacrylamide hydrogel with graphene nanoplatelets for neural cell-laden 3D bioprinting. In: Proceedings of the Annual International Conference of the IEEE Engineering in Medicine and Biology Society; EMBS 2016. 2016 October. pp. 4185-4188

[42] Park S, Park J, Sim S, et al. Enhanced differentiation of human neural stem cells into neurons on graphene. Wiley Online Library. http://onlinelibrary.wiley.com/doi/10.1002/adma.201101503/full

[43] Boyce ST, Lalley AL. Tissue engineering of skin and regenerative medicine for wound care. Burns & Trauma. 2018;**6**:4

[44] Kim BS, Lee JS, Gao G, et al. Direct 3D cell-printing of human skin with functional transwell system. Biofabrication. 2017;**9**. DOI: 10.1088/1758-5090/aa71c8

[45] Rimann M, Bono E, Annaheim H, et al. Standardized 3D bioprinting of soft tissue models with human primary cells. Journal of Laboratory Automation. 2016;**21**:496-509

[46] Lee W, Debasitis JC, Lee VK, et al. Multi-layered culture of human skin fibroblasts and keratinocytes through three-dimensional freeform fabrication. Biomaterials. 2009;**30**:1587-1595

[47] Lee V, Singh G, Trasatti JP, et al. Design and fabrication of human skin by three-dimensional bioprinting. Tissue Engineering Part C: Methods. 2014;**20**:473-484

[48] Yanez M, Rincon J, Dones A, et al. In vivo assessment of printed microvasculature in a bilayer skin graft to treat full-thickness wounds. Tissue Engineering Part A. 2015;**21**:224-233

[49] Skardal A1, Mack D, Kapetanovic E, Atala A, Jackson JD, Yoo J, et al. Bioprinted amniotic fluid-derived stem cells accelerate healing of large skin wounds. Stem Cells Translational Medicine. **1**:70-78

[50] Jang J. 3D bioprinting and in vitro cardiovascular tissue modeling. Bioengineering. 2017;**4**:71

[51] Benjamin EJ, Blaha MJ, Chiuve SE, et al. Heart disease and stroke statistics—2017 update: A report from the American Heart Association. Circulation. 2017;**135**:e146-e603

[52] Nakamura M, Kobayashi A, Takagi F, et al. Biocompatible inkjet printing technique for designed seeding of individual living cells. Tissue Engineering. 2005;**11**:1658-1666

[53] Boland T, Tao X, Damon BJ, et al. Drop-on-demand printing of cells and materials for designer tissue constructs. Materials Science and Engineering C. 2007;**27**:372-376

[54] Cui X, Boland T. Human microvasculature fabrication using

thermal inkjet printing technology. Biomaterials. 2009;**30**:6221-6227

[55] Duan B, Kapetanovic E, Hockaday LA, et al. Three-dimensional printed trileaflet valve conduits using biological hydrogels and human valve interstitial cells. Acta Biomaterialia. 2014;**10**:1836-1846

[56] Xu T, Baicu C, Aho M, et al. Fabrication and characterization of bio-engineered cardiac pseudo tissues. Biofabrication. 2009;**1**:035001

[57] Zhang YS, Arneri A, Bersini S, et al. Bioprinting 3D microfibrous scaffolds for engineering endothelialized myocardium and heart-on-a-chip. Biomaterials. 2016;**110**:45-59

[58] Amini AR, Laurencin CT, Nukavarapu SP. Bone tissue engineering: Recent advances and challenges. Critical Reviews in Biomedical Engineering. 2012;**40**:363-408

[59] Lu Y, Mapili G, Suhali G, et al. A digital micro-mirror device-based system for the microfabrication of complex, spatially patterned tissue engineering scaffolds. Journal of Biomedical Materials Research: Part A. 2006;**77**:396-405

[60] Catros S, Guillemot F, Nandakumar A, et al. Layer-by-layer tissue microfabrication supports cell proliferation in vitro and in vivo. Tissue Engineering Part C: Methods. 2012;**18**:62-70

[61] Hung BP, Naved BA, Nyberg EL, et al. Three-dimensional printing of bone extracellular matrix for craniofacial regeneration. ACS Biomaterials Science and Engineering. 2016;**2**:1806-1816

[62] Fedorovich NE, Wijnberg HM, Dhert WJA, et al. Distinct tissue formation by heterogeneous printing of osteo- and endothelial progenitor cells. Tissue Engineering Part A. 2011;**17**:2113-2121

[63] Zhou X, Zhu W, Nowicki M, et al. 3D bioprinting a cell-laden bone matrix for breast cancer metastasis study. ACS Applied Materials and Interfaces. 2016;**8**:30017-30026

[64] Hunziker EB. Articular cartilage repair: Basic science and clinical progress. A review of the current status and prospects. Osteoarthritis and Cartilage. 2002;**10**:432-463

[65] Xu T, Binder KW, Albanna MZ, et al. Hybrid printing of mechanically and biologically improved constructs for cartilage tissue engineering applications. Biofabrication. 2013;**5**. DOI: 10.1088/1758-5082/5/1/015001

[66] Schuurman W, Levett PA, Pot MW, et al. Gelatin-methacrylamide hydrogels as potential biomaterials for fabrication of tissue-engineered cartilage constructs. Macromolecular Bioscience. 2013;**13**:551-561

[67] Kundu J, Shim J-H, Jang J, et al. An additive manufacturing-based PCL-alginate-chondrocyte bioprinted scaffold for cartilage tissue engineering. Journal of Tissue Engineering and Regenerative Medicine. 2015;**9**:1286-1297

[68] Cui X, Breitenkamp K, Finn MG, et al. Direct human cartilage repair using three-dimensional bioprinting technology. Tissue Engineering Part A. 2012;**18**:1304-1312

[69] Cui X, Boland T, D'Lima DD, et al. Thermal inkjet printing in tissue engineering and regenerative medicine. Recent Patents on Drug Delivery and Formulation. 2012;**6**:149-155

[70] O'Connell CD, Di Bella C, Thompson F, et al. Development of the biopen: A handheld device for surgical printing of adipose stem cells at a chondral wound site. Biofabrication. 2016;**8**: 0

[71] Miri AK, Nieto D, Iglesias L, et al. Microfluidics-enabled multimaterial maskless stereolithographic bioprinting. Advanced Materials. 2018;**1800242**:1-9

[72] Bajaj P, Marchwiany D, Duarte C, et al. Patterned three-dimensional encapsulation of embryonic stem cells using dielectrophoresis and stereolithography. Advanced Healthcare Materials. 2013;**2**:450-458

[73] Carter S-SD, Costa PF, Vaquette C, et al. Additive biomanufacturing: An advanced approach for periodontal tissue regeneration. Annals of Biomedical Engineering. 2017;**45**:12-22

[74] Liu Y, Zheng Y, Ding G, et al. Periodontal ligament stem cell-mediated treatment for periodontitis in miniature swine. Stem Cells. 2008;**26**:1065-1073

[75] Zhu W, Liang M. Periodontal ligament stem cells: Current status, concerns, and future prospects. Stem Cells International. 2015. DOI: 10.1155/2015/972313

[76] Ma Y, Ji Y, Huang G, et al. Bioprinting 3D cell-laden hydrogel microarray for screening human periodontal ligament stem cell response to extracellular matrix. Biofabrication. 2015;**7**:44105

[77] Kim SE, Shim KM, Jang K, et al. Three-dimensional printing-based reconstruction of a maxillary bone defect in a dog following tumor removal. In Vivo. 2018;**32**:63-70

[78] Ludwig PE, Huff TJ, Zuniga JM. The potential role of bioengineering and three-dimensional printing in curing global corneal blindness. Journal of Tissue Engineering. 2018;**9**:204173141876986

[79] Whitcher JP, Srinivasan M, Upadhyay MP. Corneal blindness: A global perspective. Bulletin of the World Health Organization. 2001;**79**:214-221

[80] Gain P, Jullienne R, He Z, et al. Global survey of corneal transplantation and eye banking. JAMA Ophthalmology. 2016;**134**:167

[81] Oie Y, Nishida K. Corneal regenerative medicine. Regenerative Therapy. 2016;**5**:40-45

[82] Dekaris I. Current trends in corneal transplantation. Medical Science. 2013;**39**:35-46

[83] Isaacson A, Swioklo S, Connon CJ. 3D bioprinting of a corneal stroma equivalent. Experimental Eye Research. 2018;**173**:188-193

[84] Haagdorens M, Van Acker SI, Van Gerwen V, et al. Limbal stem cell deficiency: Current treatment options and emerging therapies. Stem Cells International. 2016;**2016**:1-22. Review article

[85] Sorkio A, Koch L, Koivusalo L, et al. Human stem cell based corneal tissue mimicking structures using laser-assisted 3D bioprinting and functional bioinks. Biomaterials. 2018;**171**:57-71

Section 2

Tissue Regeneration Concepts

Chapter 5

Application of Bone Substitutes and Its Future Prospective in Regenerative Medicine

Ujjwal Ranjan Dahiya, Sarita Mishra and Subia Bano

Abstract

Bone is a hard and dense connective tissue that supports and maintains the body structure and functions. Several factors like aging, drugs, hormonal changes, and physical activities lead to several kinds of bone injuries/fracture. To address these problems, autologous bone graft is considered an ideal material. However, limited availability and complications related to its harvesting process like donor site morbidity and pain limit the use of autologous bone graft in bone regeneration. With increasing advances in technology, several bone substitute materials such as synthetics, bioceramics, and polymers are emerging as a substitute of auto- or allogenous bone for the treatment of bone defect. These bone substitute materials should be biocompatible, bioresorbable, osteoconductive, osteoinductive, and support the ingrowth of new bone. In this chapter, we summarize the currently available bone graft and bone substitute materials including biological and bio-inorganic factors. An overview of the associated advantages, challenges, and future perspectives to clinical implication is also discussed.

Keywords: autograft, allograft, growth factors, bone graft substitute, tissue engineering

1. Introduction

Bone is a part of vertebrate skeleton. It plays a multitude of important roles in the body like imparting structural support, protection of organs, acting as a site for production of blood cells, and also as repertoire of minerals. Bones comprise differentiated cell types, blood vessels, protein, minerals, and vitamins that facilitate their growth and repair system [1]. Bones have an inner and outer layer. The hard-outer layer of bone which is called "cortical bone" is usually tough and strong, whereas the inner spongy part is called "trabecular bone" and is lighter and less dense. Each of these parts comprises different cell types, nonmineral proteinaceous matrix (osteoid), and matrix-deposited inorganic minerals. Another important concept in bone biology and understanding its transformational changes is that of modeling and remodeling. The scenario where the sites of bone formation and resorption are different surfaces of the bone is called bone modeling. This is responsible for increased length and girth of long bone, leading to skeletal development and changes. Bone remodeling on the other hand is important for maintenance of bone mass in adults by replacement of old bone tissue with new ones [2]. Several factors which affect bone, muscles, and joints are responsible for causing diseases

IntechOpen

like osteoarthritis (degenerative joint disease), rheumatoid arthritis (autoimmune disease), fibromyalgia (chronic condition of pain in bones, muscles, and tender areas with fatigue) and bone fractures. Aging plays an important role in manifestation of bone-related diseases along with other sub-factors like lifestyle, level of activity, family history, level of physical activity, drug usage, etc. However, in certain clinical situations, the natural bone repair may be too slow or inadequate; therefore, an alternative bone grafting strategy is required to address this problem.

2. Bone graft

In 1861, a surgeon from Lyon, France named Leopold Ollier was the first to describe the term bone graft (French: "greffe osseuse") in his document "Traité de la régénération des os" [3]. It was considered a surgical procedure to promote bone healing for several reasons, like injury and disease utilizing a bone transplant.

Bone graft is the alternative choice to address the problem associated with bone disease. Bone grafts are basically bone-like materials that come from living donor, post mortem donors, or artificially constructed, which are used for healing, strengthening, or improvement of bone function in disease or injury.

2.1 Properties of ideal bone graft

The ideal bone substitute material should exhibit several important properties [4, 5]. This includes:

i. *Biocompatibility*—The graft should not evoke an immune response against the implanted tissue.

ii. *Durability*—The graft should be able to maintain its shape and volume over time without loss of its structural properties.

iii. *Vascularity and angiogenesis*—Porosity is important to maintain the proper transport of nutrients and oxygen for survival of the cells and tissues. A porous structure of at least 100 μm diameter is recommended. It has also been shown that supports with multiple apertures are preferred over one large pore. Also, a polymeric, ceramic, or composite material is preferred over a metallic one as the latter might not fuse completely and dissociate after implantation. Although most scaffold (structural support) materials do not induce angiogenesis, it is an indispensable requirement to meet high blood demands in bone tissue. The initial inflammatory response activates neovascularization, yet it takes several steps to form a vascular network. The established role of proinflammatory response is their catabolic effect on bone and premature failure of bone implants. However, recently in opposition to the conventional unregulated, destructive effect of inflammatory cytokines, its regulated role aiding towards fracture healing has been suggested [6–8]. The acute inflammation initiates the recruitment of MSCs, fibroblasts, and osteoprogenitor cells to promote bone regeneration [9]. The early phase of bone injury includes the inflammation phase where IL-1, prostaglandins, TNF-α, and other proinflammatory molecules recruit inflammatory cells and induce angiogenesis. IL-6 triggers angiogenesis by stimulating release of VEGF [8]. A recent study by Anghelescu et al. shows tissue formation exceeding the catabolic action of inflammation and it might be limited to specific allografts only.

iv. *Bio-absorbability*—Another important feature of a substitute material is that it should get absorbed and provide the space for new bone formation. However, the duration of availability largely depends upon the site of surgery. The spinal fusion requires the material to be degraded after 9 months or longer while in skull or maxillofacial bone the required time period is around 3–6 months.

v. *Cost efficiency and availability*—The material should be reasonable to be purchased and used.

Besides the above-mentioned characteristics, the ideal bone substitutes must strengthen bone healing by following the mechanism which involves osteogenesis, osteoinduction, and osteoconduction [10], which are mentioned here:

vi. *Osteogenesis* (acts as the origin or source)—The process where cellular elements from the host or graft donor provide the material to synthesize new bone at the graft site is called osteogenesis. Various autografts and stem cell transplants fall under this category.

vii. *Osteoinduction* (acts as initiator)—The process resulting in differentiation of recruited mesenchymal stem cells into chondroblast and osteoblast, which forms the new bone, is called osteoinduction. It is regulated essentially by various factors like bone morphogenetic proteins (BMP)-2, -4, -7, platelet-derived growth factor (PDGF), parathyroid hormone (PTH), etc.

viii. *Osteoconduction* (provides direction)—A degradable support or scaffold material which provides the surface for the production of new bone. It also exhibits osteoconductive properties. These may include metals and synthetic polymers, and most frequently used osteoconductive materials include calcium phosphate ceramics such as hydroxyapatite, calcium sulfate, and bioactive glass-ceramics.

Most widely used bone grafts (**Figure 1**) are mentioned below in detail.

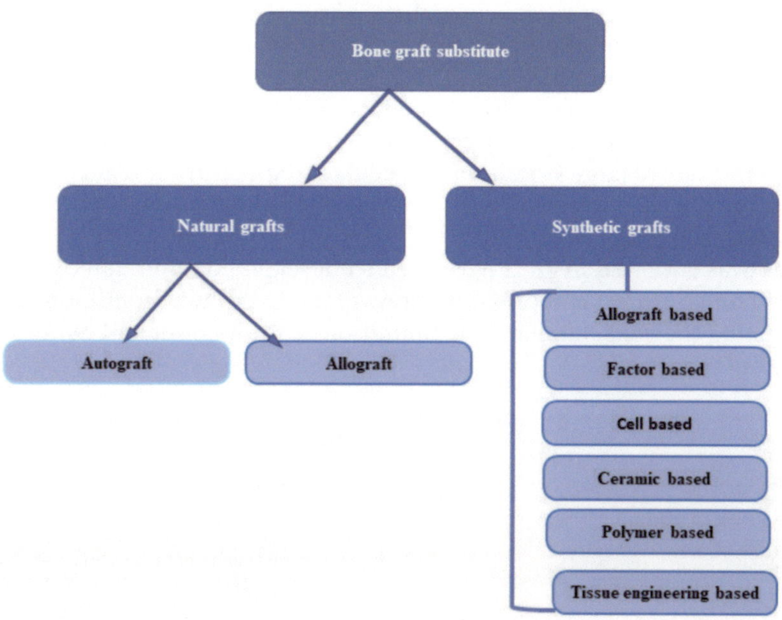

Figure 1.
Classification of bone grafting substitutes.

3. Autograft

The autologous bone grafting as the name suggests 'auto' (self) involves taking up tissue from one anatomical location and transplanting it to another in the same patient. The method dates back to 1821 when Walther repaired the holes in a patient's skull using the original bone plug [11, 12]. The method is still considered to be a 'gold standard 'as there is no immunogenic response against patient's own tissue [13]. Thus, no graft rejection or histocompatibility issues are conferred. Moreover, it manifests the properties of an ideal bone graft, that is, osteogenesis, osteoinduction, and osteoconduction. However, it is not a straightforward path to follow and comes with a set of limitations associated with it. This includes an additional operative pain at donor site, increased blood loss, and also possible injury to nearby blood vessels. Besides, the inappropriate availability of tissue amount especially from infants and older patients results in added trouble [14, 15].

3.1 Autograft-based substitute

It is considered as a gold standard to treat bone defects due to their established osteoconductive and osteoinductive properties with reduced histocompatibility issue. Although it also has inherent drawbacks associated with issues of high post-operative pain, donor site morbidity, muscle weakness, infection, high cost, and longer recovery periods limit its application.

Some of the common advantages and disadvantages associated with autograft are listed in **Table 1**.

3.2 Types of autografting commonly used are

3.2.1 Cancellous autograft

This is the most common procedure among autografts and has shown success for various purposes majorly nonunions. It starts with hematoma formation resulting in recruitment of mesenchymal stem cells (MSCs), while simultaneously, necrotic graft is eliminated by macrophages [16, 17]. Neovascularization also takes place along with this, and finally, osteoid produced by osteoblasts lining the dead trabeculae forms new bone through mineralization, which takes a period ranging from 6 to 12 months post-operation [18]. The most common donor site is the iliac crest due to large surface of trabecular architecture and availability of growth factors [19, 20]. Recent report showed superior osseous bridging after bilateral tibial tuberosity advancement (TTA) over no graft in dogs [21]. Earlier, bone union was observed in graft harvested from scaphoid nonunion distal radius or iliac crest and headless compression screw to treat scaphoid nonunion [22]. Although being the gold standard, the limitation of the amount and donor site morbidity hinders its use.

Advantages	Limitations
Osteogenic	Donor site pain
Osteoconductive	Increased blood loss
Osteoinductive	Inappropriate amount of tissue availability
No graft rejection	Increased risk of nerve injury

Table 1.
Common advantages and disadvantages associated with an autograft.

3.2.2 Cortical autograft

It shows creeping substitution, that is, deposition of new bone along with slow resorption of the graft. The process provides distinguished structural support. This graft also faces limited supply of osteoblasts and revascularization is impeded too.

3.2.3 Vascularised autograft

This type of graft promotes fracture healing and minimizes loss of bone strength after post-implantation. However, this graft also faces technical challenges for the preservation of graft's osteocytes and osteoprogenitor cells.

4. Allograft

The second category of natural bone grafts is allograft, where tissue from another individual (donor) has to be taken for grafting. The term allograft ('allo' (other)) means harvesting tissue from different individuals to transplant in the subject. Bone harvested from cadavers in different bone banks is the major source of different allografts. Apart from the above-mentioned category, freeze dried (lyophilized) and frozen bones are also used in spine graft surgeries. Allografts are considered more advantageous as they are considered less stressful to the patient. However, they are also plagued with problems of availability and immune rejection. The first report of allograft usage can be dated back to 1800s, when Macewen reported the grafting of tibial fragments to a child [23]. Experimentation reported by Bauer in the beginning of twentieth century marks the starting of the bone bank concept, where he stored bone tissues for weeks in refrigerated condition and then used them for implantation in dogs. It was also established that chilling and boiling of bone tissues before using them as allografts led to destruction of their endogenous proteins and other factors resulting in poor and slow recovery [24]. The necessities posed by the devastation of World War II resulted in advancement and growth in the bone banking approach, with refinement of methods like autoclaving, freeze-drying, irradiating, demineralizing, and chemical treatment of the bone [25]. The advancement made during the war period continued with focus on civilian need and further refinement of the banking approach [23]. Pros and cons of allografts can be summarized in **Table 2**.

Advantages	Disadvantages
Less chance of donor morbidity	Chances of disease transmission
No size limitation	Possibility of host immune response
Less surgical intervention	High cost
Cosmetically better results	Delay in incorporation
Reduce period for rehabilitation	Local bone resorption

Table 2.
Common advantages and disadvantages associated with an allograft.

4.1 Types of allografting commonly used are

4.1.1 Cancellous bone graft

This is primarily an osteoconductive graft, devoid of growth factors (not osteoinductive), and like cancellous autograft, it also provides less mechanical support [26]. It is prepared in small cubicle form, thus sometimes called 'croutons' and is now marketed in varied sizes as chips or crushed form. It involves an initial cascade of inflammatory events leading to formation of a fibrous tissue around the graft, preventing complete integration of the graft. The survival and efficiency depend a lot on processing and storage conditions of the grafts. Freeze drying is the most common method of preservation of cancellous graft, which may result in destruction of osteoprogenitor cells and osteoinductive factors [27, 28].

4.1.2 Cortical bone graft

This shows a creeping substitution like its autograft counterpart, thus leading to bone healing in spite of an initial inflammatory response. It also provides structural support and thus can be utilized for load-bearing areas [10].

Both allograft and autograft have their own advantages and disadvantages. However, research is being carried out to search the materials which can be used as a substitute to replace the function of auto- and allograft. These substitutes possess the properties of one's own bone, which stimulate bone healing and provide a strong and biologically compatible environment for the growth of new bone.

5. Bone graft substitute

The bone graft substitute is basically a synthetic inorganic or biologically organic combination, which guides to stimulate bone healing and fill bone defects. A good bone grafting substitute should have low immunogenicity and higher biocompatibility, and at the same time, it should be able to mold itself according to grafting needs. Bone grafts find application and are primarily used for linkage and splintage and help in promoting ontogenesis. In the case of linkage, bone grafts are utilized for filling cavities or defects in bone, replacement of crushed bone, and arthrodesis. While splintage-related grafting protocols are utilized for nonunion bone deformities and arthrodesis [29]. An understanding of bone formation, process of osteogenesis, osteoinduction, and osteoconduction is a prerequisite for appreciating the biology of graft substitution. The type of graft used and its physicochemical properties play an important role in its success. On a global scale, around 2.2 million graft substitutions are performed annually with 9 out of 10 falling under the category of allograft or autografts. This bone grafting procedure typically follows the multistep cascade:

- Accumulation of inflammatory cells takes place due to induction by exogenous graft material, which is followed by chemotaxis reaction by mesenchymal stem cells. After that, the host cells differentiate into chondroblasts and osteoblasts with the stimulation of several osteoinductive factors [30].

- The osteoinductive matrix serves as a scaffold for supporting progenitor cells and osteoblasts. It also provides a porous structure, which helps in migration of new cells.

- The osteoblast cells start the formation of new bone under suitable conditions.

- Lastly, the osteoinductive proteins start stimulating the differentiation of osteoblasts. This leads to initialization of bone graft revascularization and necrotic graft resorption. As a result, bone production followed by bone remodeling started from the osteoblasts on the graft's three-dimensional structure [31]. An outline of bone graft substitute process is shown in **Figure 2**.

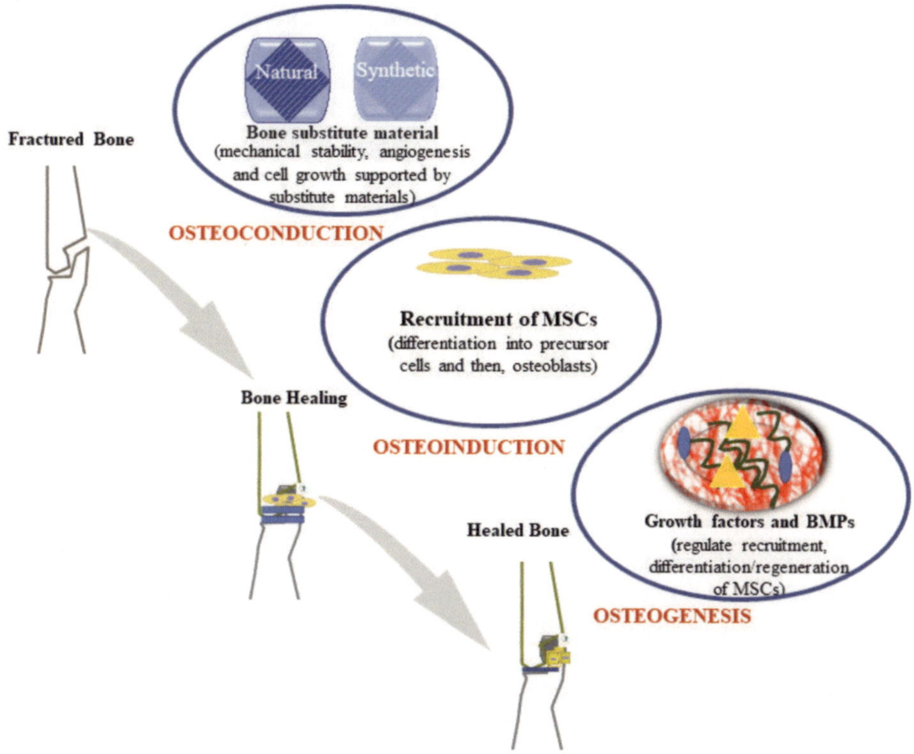

Figure 2.
Schematic representation shows the process of bone graft substitues.

Broadly, the substitute materials can be categorized as natural and synthetic grafts. Under the natural grafting substitutes fall autografts and allografts.

5.1 Natural substitute material

5.1.1 Factor-based substitutes

Factor-based bone grafts include natural and recombinant growth factors and hormones, which are generally used alone and/or together with other materials. The bone morphogenetic proteins (BMPs), platelet-derived growth factor (PDGF), parathyroid hormone (PTH), transforming growth factor-beta proteins (TGF-β), and many more come under this category. This group too does not come without associated disadvantages, such as protein instability, risk of uncontrolled proliferation or cancer, and high cost.

5.1.1.1 Bone morphogenic proteins

Nearly 30 human proteins fall under the BMP name and this group comprises the largest division of transforming growth factor (TGF)-β family of ligands. Since the last 40 years, the significance of BMP signaling in skeletal development has received immense approval [32]. It was first identified as a substance in extracellular matrix of bone, which stimulates new bone formation by Marshall R. Urist, and later named BMPs. In the United States, in 2002, the Food and Drug Administration (FDA) approved the first bone graft substitute, which was the recombinant human (rh) BMP-2 for single-level anterior lumbar interbody fusion (ALIF) within a titanium cage of implant grade. Subsequently, the increased use of BMP-2 for various spine fusion injuries as an alternative to autograft acted to be a savior as bone substitute. Despite that, starting from 2006, independent research groups started reporting complications associated with the use of rhBMP-2. The side effects ranged from seroma formation [33, 34], ectopic bone growth [35–37], osteolysis [38, 39], and increased risk of cancer [40, 41]. This led to a systematic review and meta-analysis, which was reported in Annals of Internal Medicine in June 2013. It was conducted as joint work of Medtronic and Yale University Open Data Access Project (YODA), and reported biased results in earlier industry-funded publications [42, 43]. Now, the use of rhBMP-2 (INFUSE) is limited to oblique lateral interbody fusion (OLIF) and anterior lumbar interbody fusion (ALIF) upon nonavailability of autograft. Also, BMP-7 or human osteogenic peptide-1 (OP-1) was given an FDA Humanitarian Device Exemption in 2004 [44, 45], but later studies [46] ultimately led to the FDA rejection of Pre-Market Approval of OP-1 in April 2009.

5.1.1.2 Platelet-derived growth factor (PDGF)

Platelets, monocytes, and endothelial cells are responsible for production of PDGF. They recruit various inflammatory cells at the fracture site resulting in increased cellular proliferation and collagen deposition. The potential use of PDGF has been suggested in the treatment of osteoporosis and bone healing. It is thought that recombinant PDGF-BB form assists to release the pericytes from the abluminal side to release free and activated MSCs [47, 48].

5.1.1.3 Parathyroid hormone

It is produced by parathyroid glands and is involved in calcium and bone metabolism. It is suspected that it might accelerate fracture healing and might reduce overall fracture risk [49, 50]. According to reports, PTH can exert anabolic as well as catabolic effect on bone metabolism. This feature depends extensively on the duration of administration wherein continuous administration results in bone resorption, while intermittent administration leads to increased bone deposition and induction of IGF-1 [51, 52]. Thus, it is speculated as a potential agent for osteoporosis treatments.

5.1.1.4 Insulin-like growth factor (iLGF)

The recombinant human insulin-like growth factor-1 (rhIGF-1) is also called mecasermin (International Nonproprietary Name (INN)). Children who are very short of their age are the target group of the product [53] and are FDA approved for the mentioned condition called severe primary Insulin IGF-1 deficiency under the trade name of INCRELEX. The FDA approval (in Aug, 2005) was based on five clinical trials of the drug (subcutaneously injected), of which four were open-label studies and one was double-blinded and placebo-controlled. The integrated results, followed up each year for 8 years, showed a considerable increase in height velocity (HV) with reference to pre-treatment [54]. HV is calculated between consecutive annual study visits as the difference in height, divided by the difference in age [55].

5.2 Synthetic grafts

Synthetic materials that are used for grafting should also possess the properties of biocompatibility, bioresorbability, osteoconductivity, osteoinductivity, cost-effectiveness, and the ease of use to make an ideal grafting material. It should also possess the similar features of cortical or cancellous bone which include toughness, modulus of elasticity, and compressive strength. With all these specificities and different mechanism of actions, a large number of synthetic bone alternatives have emerged which can be used for orthopedic applications.

5.2.1 Ceramic-based bone grafts

Ceramic-based grafts are ionically bonded inorganic preparations, which can be described as a family of materials with a wide variety of composition, porosity, manufacturing, and structure. Calcium sulfate, hydroxyapatite, tricalcium phosphate, bioactive glasses (silicon dioxide, calcium oxide, phosphorous pentoxide, and sodium oxide), and synthetic hydroxyapatite come under this category. Not only this group of materials varies in terms of material properties, but also differs in biodegradability, mechanistic strength, and binding. Glass ionomers have been used to seal defects in the skull, sinus augmentations, and otorhinolaryngologic surgeons for auditory ossicular reconstructions [56]. They are also used for the application of orbital implants, ossicular replacements, and prosthetic joint linings [57]. Calcium phosphate does not show any osteogenic properties; however, this material is used with the combination of hydroxyapatite and tricalcium phosphate and calcium carbonate and monocalcium phosphate monohydrate. This mixture is used as an injectable into the bone defect site, to harden. This product undergoes long-term remodeling, and the graft is eventually replaced with the in-growth of new bone [58, 59]. Hydroxyapatite-based materials are used for coat implants because of their great osteo-integrative capabilities [60, 61].

Over all, ceramic-based bone grafts are specifically found useful in iliac crest bone grafting and as a bone graft extender in lumbar spine fusion procedures [62]. The special feature of ceramic-based graft is their porosity, which is helpful in adhesion of mesenchymal cells, and later gets differentiated into osteoblasts. Further, this group of grafts is superior as the benefits of its different constituents can be availed simultaneously leading to better bone regeneration [63]. Some of the commercially available synthetic bone graft substitutes are mentioned in **Table 3**.

5.3 Polymer-based bone grafts

Polymer-based grafts can be broadly grouped as synthetic or natural polymer-based grafts and can further be classified as biodegradable and nonbiodegradable substitutes. This category of grafting materials is specifically utilized for dental implants, as they are found helpful in restoring of edentulous site in the lost tooth. In the case of long bone-related congenital defects, or cases requiring bone segment replacements after invasion of malignant tissues, polymer-based grafts (vascular fibulas) are found helpful in restoring skeletal integrity. Other polymers like chitosan, collagen, gelatin, and poly lactic acid (PLA) are reported to exhibit improved bone regeneration capabilities both *in vitro* and *in vivo* when combined with hyaluronic acid [64]. Another advantage shown by these materials is that of better bone-matrix interference since they also act as bio-mimetic, thus helping in deposition, precipitation, and enhancing formation of calcium phosphate [65]. The type of commercially available polymer-based bone graft substitute is mentioned in **Table 4**.

5.4 Tissue engineering and cell-based bone graft

The interdisciplinary field of tissue engineering seeks to combine the bone graft substitutes to stimulatory effects of growth factors, (bone morphogenetic proteins and osteogenic proteins) to provide structural support and promote more rapid bone growth and healing. The prime aim of the bone tissue engineering approach is to

Commercial product (name)	Substitute materials	Properties	Applications
Osteograf	Ceramic	Osteoconductive, limited osteoinductive when mixed with bone marrow	Bone void filler
NovaBone	Bioactive glass	Osteoconductive, limited osteoinductive when mixed with bone marrow	Filling surgical or traumatic bone gaps
Osteosat	Surgical grade calcium phosphate	Osteoconductive and bioresorbable	Hip and knee joint repair
Calceon 6	Calcium sulfate	Osteoconductive and bioresorbable	Bone void filler; provides strength
Norian	Monocalcium phosphate, tricalcium phosphate, and calcium carbonate	Good compressive strength	Skull bone defect; injectable paste, craniofacial reconstructions
Hard tissue-replacement (HTR)	Poly methyl methacrylate (PMMA)	Good strength, durable, and surface osteoconductive	Craniofacial reconstruction
Alpha BSM	Calcium phosphate cement	Good compressive strength	Dental application for bone and cartilage defects
Mimix	Synthetic hydroxyapatite tetra-tricalcium phosphate	Good compressive strength	Cranial defects
ELIZ (Kyeron)	Composed of (40%) β-tricalcium phosphate and of (60%) hydroxyapatite	Ultrahigh porosity, biocompatible, and osteoconductive	It has been successfully implanted in more than 1200 patients without any side-effects.
OSIQ (Kyeron)	Fully synthetic ultrapure nano-hydroxyapatite	Ready to use, injectable, and biodegradable	Filling or reconstruction of small and medium bone defects
AXOZ QS (Kyeron)	Resorbable phosphocalcic compounds and a polymer	Injectable and fully resorbs	Supports bone growth
COLLAPAT II (Symatese)	Composed of a collagen structure in which ceramised hydroxyapatite granules are dispersed	Strong hemostatic power, completely resorbable in a few weeks, and osteoconductive	Induces bone substance replacement in maxillofacial surgery and odontostomatology
CopiOs (Zimmer Biomet) Bone Void Filler	Calcium phosphate, dibasic (DICAL), and highly purified Type I bovine collagen	DICAL provides significantly more calcium and phosphate ions at equilibrium than either β-TCP or HA	CopiOs paste acts as an osteoconductive scaffold for the growth of new bone

Table 3.
List of some commercially available synthetic materials and their applications.

guide and enhance osteogenic differentiation of stem cells into 3D constructs, so that later it can be engineered successfully into applicable bone constructs. The limitation in availability of both allografts and autografts can be addressed to a great extent by tissue engineered bones or healing of fracture critical defects. The tissue engineering approach for bone grafts differ from other approaches, in the fact that in earlier case,

Commercial product (name)	Substitute material	Properties	Applications
Cortoss	Polymer system with reinforcing particle bioactive glass	Forms biological interface	Augmentation of screws in osteoporotic bone (hip, spine, etc.)
Open porosity polylactic acid polymer (OPLA)	Polylactic acid	Osteoconductive and bioresorbable	Articular cartilage regeneration
Collagraft	Mixture of tricalcium phosphate, bovine collagen, and hydroxyapatite	Bioresorbable and osteoconductive	Use for the treatment of long bone fracture and void filling
DynaGraft	Demineralized bone matrix	Heat sensitive copolymer, injectable gel, limited osteo-induction	Dental bone graft substitute
MedPor	Porous polyethylene	Higher porosity	Orbital reconstruction and facial contouring
Collapro/matrix	Human collagen in lyophilized strip	Lack of immunogenic property	Use in development
Healos	Hyaluronic acid-coated collagen sponge	Osseo-inductive property	Replacement of autograft/autograft extender for spinal fusion
Immix	PGA/PLA polymer to be produced in chip, flex forms	Provides structural support	Bone graft extender
OsteoScaf (Bonetec)	Macroporous poly(lactide-co-glycolide)/calcium phosphate (PLGA/ CaP) foam matrices	Fully resorbable, osteoconductive, and mechanically robust	Heal tissue defects

Table 4.
List of some commercially available polymer-based graft materials and their applications.

integration of engineered bone with patients takes place, thus providing specific and more resilient treatment for condition [66]. Further tissue engineering is more convenient in the case of bone regeneration therapies, as osteogenic differentiation can be achieved through multiple stem cell types [67]. The type of scaffolds and cells used for tissue engineering graft is discussed in more detail.

5.4.1 Scaffolds

One important requirement for bone tissue engineering is the scaffold, which helps in the migration, proliferation, and differentiation of osteogenic cells promoting new bone formation and regeneration [68]. In this regard, it is required that the scaffold must be stable, biocompatible, and biodegradable and should be porous and permeable for cell seeding, nutrient transport, tissue ingrowth, and vascularization. It should also be osteoconductive, osteoinductive, and osseo-integrative in nature. In clinical settings, a wide range of synthetic and natural scaffolds have been explored for bone repair and bone tissue engineering. Broadly, these materials can be categorized as composites, ceramics, and polymers, with each having specific properties and limitations. Type I collagen polymer matrix serves as a good natural material but suffers from low mechanical modulus.

Market name	Allograft type	Form of/additive with	Carrier used
Grafton	DBM	Gel, putty, and flexible sheets	Glycerol carrier
Opteform	DBM	Cortical bone chips	Gelatin carrier
Osteofil	DBM	Injectable bone paste	Collagen-based hydrogel matrix
Dynagraft	DBM	Syringe	Pleuronic reverse phase copolymer
Orthoblast	DBM	Cancellous bone	Bioresorbable reverse phase copolymer

Table 5.
Marketed DBM-based bone graft substitutes.

One important group of scaffold, which has been extensively used for tissue engineering, is demineralized bone matrix (DBM). DBM is typically produced from allograft bone by the acid extraction method (known as decalcification). Based on manufacturing techniques, DBM may be presented in different forms like sponges, freeze-dried powder, gel, paste, injectable putty, or strips. It is basically a decalcified allograft retaining collagen and noncollagenous proteins [69]. It is osteoconductive as well as osteoinductive but has no osteogenic property due to its processing. Recently, it is being used in conjunction with cancellous/cortical bone chips. The unlimited availability and reduced immune response owing to acidic demineralization give it superiority while it is generally used as a bone graft additive in spine fusions [70]. The types of DBM bone graft substitute which is commercially available are mentioned in **Table 5**.

5.4.2 Cell type involved in bone tissue engineering

A variety of *in vitro* culture protocols are used in bone tissue engineering, which requires use of growth factors and cytokines along with specialized dynamic bioreactors. It also requires a large scale of bone extracellular matrix-producing cells, primarily the MSCs. In this regard, the adult bone marrow stem cells (BMSCs) are the cells of choice for tissue engineering. The quantity of the isolated stem cells varies between patients and is indirectly proportional to patient's age [71, 72]. Also, the isolated cells should have high biosynthetic activity, expression of osteogenic markers, and right cell phenotype. Another source for cell-based bone tissue engineering is the adipose tissue which consists of adipose stem cells (ASCs). They undergo differentiation into several cell lineages such as osteogenic, chondrogenic, and endothelial. Comparative studies of ASCs and BMSCs have shown that they share their surface antigen expression pattern involving CD44, CD71, CD90, and CD105 [73]. However, there are differences too, such as the BMSCs mark the absence of hematopoietic and endothelial lineage markers [74–76]. Researchers are currently investigating the ability of other multipotent cells such as periosteum, umbilical cord, cord blood, and fetal tissues to be used for bone tissue engineering [77].

5.4.3 In-vitro culturing

In this approach, bone marrow is harvested from the patient, followed by *in vitro* culturing and seeding them on scaffolds prior to implantation into the same patient for tissue regeneration. The bone repair process required many cell types. These cell types are involved in many inflammatory and angiogenic responses and play an important role in development of the bone formation mechanisms. They release cytokines and various growth factors like PDGF, BMPs, VEGF, and interleukins that attract and simulate the mesenchymal stem cells (MSCs), which are directly involved in bone repair. However, cell sourcing for bone regeneration is still a critical issue for cell-based therapies that need to be addressed. Also, for the sake of achieving homogenous growth in 3D environment, a new bioreactor cultivation system with mass transport capabilities has been explored [78].

5.4.4 *In-vivo studies on bone graft applications*

Multiple reports have been published showing not only the application of bone tissue engineering in the case of ectopic bone formation (nonbone environment) [79] but also in orthotropic bone formation (bone environment) [80]. In 1991, Caplan first reported the usage of autologous stem cells for fast and specific skeletal tissue repair. Mesenchymal stem cells (MSCs) or mesenchymal colony-forming cells (CFU-F) were identified as cell type present in the bone marrow, which have the capability to differentiate into different cell types of bone [81]. In 2004, first time successful transplantation in humans was reported, where lamellar bone was regenerated and repaired by infusing periosteum-derived osteoblasts for 90 days [82]. Thus, bone marrow has been recognized as a source of MSCs and other osteogenic cells, with relative ease of cell collection using the aspiration method [83]. Ever since the identification and recognition of MSCs for repair and engineering of mesenchymal tissues like cartilage and bone, new cell types and putative strategies have been explored vastly. Among newly developed cell types for bone tissue engineering, deciduous dental pulp-derived stem cells and adipose derived cells were the prominent one showing osteogenic potential [84]. Successful tissue engineering-based bone regeneration depends on multiple factors like the number of healthy osteogenic cells transplanted, the use of proper scaffold for seeding the cells at the site of regeneration, enough vascularization of repair site, and osteogenic differentiation factors. The report provided by Coventry and Tapper [62] has proved successful regeneration of ectopic bones in rodent models, fulfilling all the above-mentioned requirements using ceramic based scaffolds [85]. Further bone marrow aspirate and plasmid-rich plasma approaches have been explored for facilitating effective tissue engineering-based grafts.

In addition to these materials, research is still continuing to modify the products for the production of graft materials, which possess all the properties of ideal bone graft. A multidisciplinary approach will be required to improve implanted cell survival on biomaterial substrate with addition of cytokines and other growth factors for prompt new bone ingrowths. This will yield a bony union that resembles natural form and structure. However, the unavailability of widely accepted specific guidelines, standardizing minimum standard for bone grafts, is one of the limiting factors. Also, the size of *in vitro* grown tissue grafts is often found to be small, in the case of critical size defects.

6. Conclusion and future perspective

The challenge of bone loss and other musculoskeletal complications during surgeries is well recognized, and to address the same issue, different grafting substitutes have emerged. These bone regeneration grafts can be broadly categorized as natural and synthetic grafting materials. Autograft and allograft comprise natural grafting substitutes and between these, allografts are gaining ground and more acceptability with time. There have been revolutionary changes during the past decade, which bring allograft as a first alternative after autograft, which were subsequently interchanged and/or replaced by demineralized bone matrix in certain circumstances [83]. The selection of ideal bone graft substitute materials is a difficult task and often lacks substantial scientific backing. However, it is a very fast and broad field of investigation. Current trends remarkably increase the use of synthetic bone graft as alternatives. The introduction of cost-effective, biologically improved synthetic materials that owe the property of ideal bone graft addresses the present clinical needs for the optimal treatment. It is not mandatory to use these synthetic materials alone for reconstructive procedures. However, when it used in the right situations and in combination with autologous, allograft, or other synthetics, in combination with growth factors, they give potentially more desirable results.

With the advancement of biomechanical research, an interdisciplinary field of tissue engineering has emerged. Limitations associated with the autologous bone grafting method can be addressed to a great extent with the help of bone tissue engineering. However, selection of optimized combinations of cells, synthetic materials (scaffolds) and factors will remain a challenging and complex process. One of them is the selection of materials and cells. This may include the materials, which possess appropriate mechanical properties, degradation rates, and chemical functionality; likewise, for stem cells, it should be isolated from patient-specific cells from the appropriate lineage and directed down on a scaffold construct to heal the proper bone tissues. Cheaper and safer alternatives will probably emerge with a better understanding of the inherent ability of material to induce bone formation after bone graft implantation.

Thus, these developments must also be nurtured and monitored by the group of clinicians and researchers with the knowledge of medical necessity, basic biological principles, and commercial practicality. This may be used for the production of versatile, and easy to implement, allowing for bioengineered bone grafts to more quickly make the leap from bench to bedside. This will help to improve the quality of life of pediatric and adult patients suffering from bone disease and/or disorder. This may be helpful for bridging the gap between bone tissue engineering research and its clinical implications.

Author details

Ujjwal Ranjan Dahiya[1†], Sarita Mishra[1†] and Subia Bano[2*]

1 CSIR-Institute of Genomics and Integrative Biology, New Delhi, India

2 ELVESYS Microfluidic Innovation Centre, Paris, France

*Address all correspondence to: subia.bioinfo@gmail.com

† These authors contributed equally

References

[1] Kronenberg HM. Developmental regulation of the growth plate. Nature. 2003;**423**(6937):332-336

[2] Kubota T, Michigami T, Sakaguchi N, Kokubu C, Suzuki A, Namba N, et al. Lrp6 hypomorphic mutation affects bone mass through bone resorption in mice and impairs interaction with Mesd. Journal of Bone and Mineral Research. 2008;**23**:1661-1671

[3] Donati D, Zolezzi C, Tomba P, Viganò A. Bone grafting: Historical and conceptual review, starting with an old manuscript by Vittorio Putti. Acta Orthopaedica. 2007;**78**(1):19-25

[4] Habibovic P, de Groot K. Osteoinductive biomaterials—Properties and relevance in bone repair. Journal of Tissue Engineering and Regenerative Medicine. 2007;**1**:25-32

[5] Yi H, Ur Rehman F, Zhao C, Liu B, He N. Recent advances in nano scaffolds for bone repair. Bone Research. 2016;**4**:16050

[6] Anghelescu V, Neculae I, Dincă O, Vlădan C, Socoliuc C, Cioplea M, et al. Inflammatory-driven angiogenesis in bone augmentation with bovine hydroxyapatite, B-tricalcium phosphate, and bioglasses: A comparative study. Journal of Immunology Research. 2018;**2018**:9349207

[7] Einhorn T, Gerstenfeld L. Fracture healing: Mechanisms and interventions. Nature Reviews Rheumatology. 2014;**11**(1):45-54

[8] Mountziaris P, Mikos A. Modulation of the inflammatory response for enhanced bone tissue regeneration. Tissue Engineering Part B: Reviews. 2008;**14**(2):179-186

[9] Loi F, Córdova L, Pajarinen J, Lin T, Yao Z, Goodman S. Inflammation, fracture and bone repair. Bone. 2016;**86**:119-130

[10] Roberts TT, Rosenbaum AJ. Bone grafts, bone substitutes and orthobiologics the bridge between basic science and clinical advancements in fracture healing. Organogenesis. 2012;**8**(4):114-124

[11] Sanan A, Haines SJ. Repairing holes in the head: A history of cranioplasty. Neurosurgery. 1997;**41**:767-775

[12] Laurencin C, Khan Y, Veronick J. Bone graft substitutes: past, present, and future. Bone Graft Substitutes and Bone Regenerative Engineering. 2nd edition. West Conshohocken, PA: ASTM International; 2014. 1-9

[13] Dimitriou R, Jones E, McGonagle D, Giannoudis P. Bone 7regeneration: Current concepts and future directions. BMC Medicine. 2011;**9**:66

[14] Greenwald A, Boden S, Goldberg V, Khan Y, Laurencin C, Rosier R. Bone-graft substitutes: Facts, fictions, and applications. Journal of Bone and Joint Surgery. 2001;**83**:98-103

[15] Kowalczewski C, Saul J. Biomaterials for the delivery of growth factors and other therapeutic agents in tissue engineering approaches to bone regeneration. Frontiers in Pharmacology. 2018;**9**:513

[16] Khan S, Cammisa F, Sandhu H, Diwan A, Girardi F, Lane J. The biology of bone grafting. The Journal of the American Academy of Orthopaedic Surgeons. 2005;**13**(1):77-86

[17] Older M. Bone Implant Grafting. London: Springer London; 1992

[18] Oakes D, Cabanela M. Impaction bone grafting for revision hip arthroplasty: Biology and clinical

applications. The Journal of the American Academy of Orthopaedic Surgeons. 2006;**14**(11):620-628

[19] Fini M, Giavaresi G, Torricelli P, Borsari V, Giardino R, Nicolini A, et al. Osteoporosis and biomaterial osteointegration. Biomedicine & Pharmacotherapy. 2004;**58**(9):487-493

[20] Galia CR. Moreira L. The Biology of Bone Grafts, Recent Advances in Arthroplasty. In: S. Fokter, editor. InTech; 2012. ISBN: 978-953-307-990-5. Available from: http://www.intechopen.com/books/recent-advances-in-arthroplasty/the-biology-of-bone-grafts

[21] James D, Webster N, White J, Marchevsky A, Cashmore R, Havlicek M, et al. Comparison of bone healing, as assessed by computed tomography, following tibial tuberosity advancement in dogs with and without autogenous cancellous bone grafts. New Zealand Veterinary Journal. 2017;**65**(5):270-276

[22] Kim J, Yoon J, Baek H. Corticocancellous bone graft vs cancellous bone graft for the management of unstable scaphoid nonunion. Orthopaedics & Traumatology, Surgery & Research. 2018;**104**(1):115-120

[23] Friedlaender GE, Mankin HJ, Sell KW. Osteochondral Allografts. Boston, Toronto: Little, Brown and Company; 1983. pp. 259-274

[24] Urist MR, O'Connor BT, Burwell RG. Bone Grafts, Derivatives and Substitutes. Oxford, United Kingdom: Butterworth and Heinemann; 1994

[25] Beebe K, Benevenia J, Tuy B, DePaula C, Harten R, Enneking W. Effects of a new allograft processing procedure on graft healing in a canine model: A preliminary study. Clinical Orthopaedics and Related Research. 2008;**467**(1):273-280

[26] Kang S, Han I, Hong S, Cho H, Kim W, Kim H. The MRI appearances of cancellous allograft bone chips after the excision of bone tumours. The Bone & Joint Journal. 2015;**97-B**(1):121-128

[27] Faour O, Dimitriou R, Cousins C, Giannoudis P. The use of bone graft substitutes in large cancellous voids: Any specific needs? Injury. 2011;**42**:S87-S90

[28] Atasoy A, Kose GT. Biology of cancellous bone graft materials and their usage for bone regeneration. JSM Biotechnology & Biomedical Engineering. 2016;**3**(2):1015

[29] Hanssens L, Reginster JY. Relevance of bone mineral density, bone quality and falls in reduction of vertebral and non-vertebral fractures. Journal of Musculoskeletal & Neuronal Interactions. 2003;**3**:189-193

[30] Goldberg VM. Natural history of autografts and allografts. In: Bone Implant Grafting. 1992. DOI: 10.1007/978-1-4471-1934-0_2

[31] Zipfel GJ, Guiot BH, Fessler RG. Neurosurgical Focus. 2003;**14**(2):e8

[32] Salazar V, Gamer L, Rosen V. BMP signalling in skeletal development, disease and repair. Nature Reviews Endocrinology. 2016;**12**(4):203-221

[33] Garrett M, Kakarla U, Porter R, Sonntag V. Formation of painful seroma and edema after the use of recombinant human bone morphogenetic protein-2 in posterolateral lumbar spine fusions. Neurosurgery. 2010;**66**(6):1044-1049

[34] Scheer J, Dahdaleh N, Smith Z. Seroma observed 6 months after anterior lumbar interbody fusion that included use of recombinant bone morphogenetic protein 2. The Spine Journal. 2015;**15**(10):e33

[35] Chen NF, Smith ZA, Stiner E, Armin S, Sheikh H, Khoo LT. Symptomatic ectopic bone formation after off-label use of recombinant human bone morphogenetic protein-2 in transforaminal lumbar interbody fusion. Journal of Neurosurgery: Spine. 2010;**12**(1):40-46. The Spine Journal. 2011;**11**(1):87-87

[36] Deutsch H. High-dose bone morphogenetic protein–induced ectopic abdomen bone growth. The Spine Journal. 2010;**10**(2):e1-e4

[37] Wong D, Kumar A, Jatana S, Ghiselli G, Wong K. Neurologic impairment from ectopic bone in the lumbar canal: A potential complication of off-label PLIF/TLIF use of bone morphogenetic protein-2 (BMP-2). The Spine Journal. 2008;**8**(6):1011-1018

[38] Pradhan B, Bae H, Dawson E, Patel V, Delamarter R. Graft resorption with the use of bone morphogenetic protein: Lessons from anterior lumbar interbody fusion using femoral ring allografts and recombinant human bone morphogenetic protein-2. Spine. 2006;**31**(10):E277-E284

[39] McClellan J, Mulconrey D, Forbes R, Fullmer N. Vertebral bone resorption after transforaminal lumbar interbody fusion with bone morphogenetic protein (rhBMP-2). Journal of Spinal Disorders & Techniques. 2006;**19**(7):483-486

[40] Carragee E, Chu G, Rohatgi R, Hurwitz E, Weiner B, Yoon S, et al. Cancer risk after use of recombinant bone morphogenetic protein-2 for spinal arthrodesis. Bone & Joint Journal. 2013;**95**(17):1537-1545

[41] Bach D, Park H, Lee S. The dual role of bone morphogenetic proteins in cancer. Molecular Therapy—Oncolytics. 2018;**8**:1-13

[42] Simmonds M, Brown J, Heirs M, Higgins J, Mannion R, Rodgers M, et al. Safety and effectiveness of recombinant human bone morphogenetic protein-2 for spinal fusion. Annals of Internal Medicine. 2013;**158**(12):877

[43] Fu R, Selph S, McDonagh M, Peterson K, Tiwari A, Chou R, et al. Effectiveness and harms of recombinant human bone morphogenetic protein-2 in spine fusion: A systematic review and meta-analysis. Annals of Internal Medicine. 2013;**158**(12):890-902

[44] Grauer J, Patel T, Erulkar J, Troiano N, Panjabi M, Friedlaender G. Evaluation of OP-1 as a graft substitute for intertransverse process lumbar fusion. Spine. 2001;**26**(2):127-133

[45] Cunningham B, Shimamoto N, Sefter J, Dmitriev A, Orbegoso C, McCarthy E, et al. Osseointegration of autograft versus osteogenic protein–1 in posterolateral spinal arthrodesis. The Spine Journal. 2002;**2**(1):11-24

[46] Munns J, Park D, Singh K. Role of osteogenic protein-1/bone morphogenetic protein-7 in spinal fusion. Orthopedic Research and Reviews. 2009;**1**:11-21

[47] Graham S, Leonidou A, Lester M, Heliotis M, Mantalaris A, Tsiridis E. I nvestigating the role of PDGF as a potential drug therapy in bone formation and fracture healing. Expert Opinion on Investigational Drugs. 2009;**18**(11):1633-1654

[48] Caplan A, Correa D. PDGF in bone formation and regeneration: New insights into a novel mechanism involving MSCs. Journal of Orthopaedic Research. 2011;**29**(12):1795-1803

[49] Qin L, Raggatt L, Partridge N. Parathyroid hormone: A double-edged sword for bone metabolism. Trends in Endocrinology and Metabolism. 2004;**15**(2):60-65

[50] Wein M, Kronenberg H. Regulation of bone remodeling by parathyroid hormone. Cold Spring Harbor Perspectives in Medicine. 2018;**8**(8):a031237

[51] Kroll M. Parathyroid hormone temporal effects on bone formation and resorption. Bulletin of Mathematical Biology. 2000;**62**(1):163-188

[52] Carter P, Schipani E. The roles of parathyroid hormone and calcitonin in bone remodeling: Prospects for novel therapeutics. Endocrine, Metabolic & Immune Disorders Drug Targets. 2006;**6**(1):59-76

[53] Rosenfeld RG. The IGF system: New developments relevant to pediatric practice. In: IGF-I and IGF Binding Proteins. Basic Research and Clinical Management. Endocr Dev. Basel: Karger; 2005. pp. 1-10. DOI: 10.1159/000085716

[54] [Internet]. Accessdata.fda.gov. 2018 [cited 11 November 2018]. Available from: https://www.accessdata.fda.gov/drugsatfda_docs/label/2005/021839lbl.pdf

[55] Kelly A, Winer K, Kalkwarf H, Oberfield S, Lappe J, Gilsanz V, et al. Age-based reference ranges for annual height velocity in US children. The Journal of Clinical Endocrinology & Metabolism. 2014;**99**(6):2104-2112

[56] Moore WR, Graves SE, Bain GI. Synthetic bone graft substitutes. ANZ Journal of Surgery. 2001;**71**:354-361

[57] Constantino PD, Freidman CD. Synthetic bone graft substitutes. Otolaryngologic Clinics of North America. 1994;**27**:1037-1073

[58] Campana V, Milano G, Pagano E, Barba M, Cicione C, Salonna G, et al. Bone substitutes in orthopaedic surgery: From basic science to clinical practice. Journal of Materials Science. Materials in Medicine. 2014;**25**(10):2445-2461

[59] Wenhao W, Kelvin W, Yeungab K. Bone grafts and biomaterials substitutes for bone defect repair. Bioactive Materials. 2017;**2**:224-247

[60] Bellemans J. Osseointegration in porous coated knee arthroplasty: The influence of component coating type in sheep. Acta Orthopaedica Scandinavica. Supplementum. 1999;**288**:1-35

[61] Strnad Z, Strnad J, Povysil C, Urban K. Effect of plasma sprayed hydroxyapatite coating on the osteoconductivity of commercially pure titanium implant biomechanical testing of alloplastic PMMA cranioplasty materials. Eppley BL. The Journal of Craniofacial Surgery. 2005;**16**(1):140-143

[62] Coventry MB, Tapper EM. Pelvic instability: A consequence of removing iliac bone for grafting. The Journal of Bone and Joint Surgery. American Volume. 1972;**54**:83-101

[63] Athanasiou KA, Zhu C, Lanctot DR, Agrawal CM, Wang X. Fundamentals of biomechanics in tissue engineering of bone. Tissue Engineering. 2000;**6**(4):361-381

[64] Rodrigues CV, Serricella P, Linhares AB, Guerdes RM, Borojevic R, Rossi MA, et al. Characterization of a bovine collagen-hydroxyapatite composite scaffold for bone tissue engineering. Biomaterials. 2003;**24**(27):4987-4997

[65] Saikia KC, Bhattacharya TD, Bhuyan SK, Talukdar DJ, Saikia SP, Jitesh P. Calcium phosphate ceramics as bone graft substitutes in filling bone tumor defects. Indian Journal of Orthopaedics. 2008;**42**(2):169-172

[66] Ohgushi H, Caplan AI. Stem cell technology and bioceramics: From cell to gene engineering. Journal of Biomedical Materials Research. 1999;**48**:913-927

[67] Berry E, Brown JM, Connell M, Craven CM, Efford ND, Radjenovic A,

et al. Preliminary experience with medical applications of rapid prototyping by selective laser sintering. Medical Engineering & Physics. 1997;**19**:90-96

[68] Vunjak-Novakovic G, Goldstein SA. In: Mow VC, Huiskes R, editors. Basic Orthopaedic Biomechanics and Mechano-biology. 3rd ed. Philadelphia: Lippincott Williams and Wilkins; 2005. pp. 343-408

[69] David A, Christopher E, Miller J, Jonathan N, Grauer P. The comprehensive treatment of the aging spine. The role for biologics in the aging. Spine. 2011:384-387. ISBN: 978-1-4377-0373-3

[70] Gruskin E, Doll BA, Futrell FW, Schmitz JP, Hollinger JO. Demineralized bone matrix in bone repair: History and use. Advanced Drug Delivery Reviews. 2012;**64**(12):1063-1077

[71] Wen B, Freilich M, Kuhn L. Stem cell biology and tissue engineering in dental sciences. Bone Tissue Engineering Around Dental Implants. 1st ed. 2015:749-764. eBook ISBN: 9780123977786

[72] Barrilleaux B, Phinney DG, Prockop DJ, O'Connor KC. Review: Ex vivo engineering of living tissues with adult stem cells. Tissue Engineering. 2006;**12**:3007-3019

[73] Kern S, Eichler H, Stoeve J, Kluter H, Bieback K. Comparative analysis of mesenchymal stem cells from bone marrow, umbilical cord blood, or adipose tissue. Stem Cells. 2006;**24**:1294-1301

[74] Meinel L, Karageorgiou V, Hofmann S, et al. Engineering bone-like tissue in vitro using human bone marrow stem cells and silk scaffolds. Journal of Biomedical Materials Research. Part A. 2004;**71A**:25-34

[75] Shanti RM, Li WJ, Nesti LJ, Wang X, Tuan RS. Adult mesenchymal stem cells: Biological properties, characteristics, and applications in maxillofacial surgery. Journal of Oral and Maxillofacial Surgery. 2007;**65**:1640-1647

[76] Chamberlain G, Fox J, Ashton B, Middleton J. Concise review: Mesenchymal stem cells: Their phenotype, differentiation capacity, immunological features, and potential for homing. Stem Cells. 2007;**25**:2739-2749

[77] Sudo K, Kanno M, Miharada K, et al. Mesenchymal progenitors able to differentiate into osteogenic, chondrogenic, and/or adipogenic cells in vitro are present in most primary fibroblast-like cell populations. Stem Cells. 2007;**25**:1610-1617

[78] Grayson WL, Chao PG, Marolt D, et al. In: Translational Approaches in Tissue Engineering and Regenerative Medicine. Boston: Artech House; 2007. pp. 353-374

[79] Haynesworth SE, Goshima J, Goldberg VM, Caplan A. Characterization of cells with osteogenic potential from human marrow. Bone. 1992;**13**:81-88

[80] Yoshikawa T, Ohgushi H, Nakajima H, Yamada E, Ichijima K, Dohi Y, et al. In vivo osteogenic durability of cultured bone in porous ceramics: A novel method for autogenous bone graft substitution. Transplantation. 2000;**69**:128-134

[81] Friedenstein AJ, Chailakhyan RK, Gerasimov UV. Bone marrow osteogenic stem cells: In vitro cultivation and transplantation in diffusion chambers. Cell and Tissue Kinetics. 1987;**20**(3):263-272

[82] Schimming R, Schmelzeisen R. Tissue-engineered bone for maxillary sinus augmentation. Journal of Oral and Maxillofacial Surgery. 2004;**62**(6):724-729

[83] Wada T, Kaya M, Nagoya S, Kawaguchi S, Isu K, Yamashita T, et al. Complications associated with bone cementing for the treatment of giant cell tumors of bone. Journal of Orthopaedic Science. 2002;7(2):194-198

[84] Laino G, Graziano A, d'Aquino R, Pirozzi G, Lanza V, Valiante S, et al. An approachable human adult stem cell source for hard-tissue engineering. Journal of Cellular Physiology. 2006;**206**:693-701

[85] Goshima J, Goldberg VM, Caplan AI. Osteogenic potential of culture-expanded rat marrow cells as assayed in vivo with porous calcium phosphate ceramic. Biomaterials. 1991;**12**:253-258

Chapter 6

Biopharmaceutical Products and Biomaterials of the Amazon Region Used in Dentistry

Maria Fulgência Bandeira, Nikeila Conde, Carina Toda, Ana Regina Casaroto, Geisy Lima, Marne Vasconcellos, Leilane Bentes, Larissa Silva, Samuel Lungareze and Joyce Meira

Abstract

In dentistry, biomaterials are used in restorative procedures, such as dental restorations; in dentures; dental implants; surgical procedures; and endodontic materials. Most dental biomaterials are classified as devices, including filling materials, diagnostic aids, cements, bonding agents, and implants, in addition to mouthwashes. In the field of health, the use of natural products for dental biomaterials and curing diseases has always emphasized, rather than depending on the conventional allopathic medicine. Brazil has an advantage in this market, because it has the greatest biodiversity in the world, especially in the Amazonian Region, and a genetic heritage of great potential for the development of new herbal products, especially in dentistry. Given the growth of products derived from medicinal plants in Brazil, it was necessary to implement a statute that covered the requirements for all medicines and biomaterials to ensure the quality, efficacy, and safety of these products. Thus, researches in dentistry have been developed with the aim of searching for new bioactive principles for the formulation of drugs with different types of applications, capable of acting in both preventive strategies and curative treatments. This has encouraged the use of phytotherapic agents such as *Copaifera multijuga, Apis mellifera* (propolis), and *Libidibia ferrea*.

Keywords: medicinal plants, *Copaifera multijuga*, *Apis mellifera*, *Libidibia ferrea*, dentin surface, orabase, mouthwash

1. Introduction

According to the American National Institute of Health (NIH), biomaterial is defined as "any substance or combination of substances, other than drugs, synthetic or natural in origin, which can be used for any period of time, which augments or replaces partially or totally any tissue, organ or function of the body, in order to maintain or improve the quality of life of the individual" [1]. The advances led to a pronounced increase in the range of use and efficacy of biomaterials over time. Thus, biomaterials have become critical components used in many industries, including medical devices, dental restoratives, and drug delivery, and are increasingly being used in technological applications such as in vitro diagnostics [2, 3].

These materials must be biocompatible at the material-tissue interface: "ability of a biomaterial to perform its desired function with respect to a medical therapy, without eliciting any undesirable local or systemic effects in the recipient or beneficiary of that therapy, but generating the most appropriate beneficial cellular or tissue response to that specific situation" [1]. In dentistry, biomaterials are used in restorative procedures, such as dental restorations; dentures; dental implants; surgical procedures; and endodontic materials. Most dental biomaterials are classified as devices, including filling materials, diagnostic aids, cements, bonding agents, and implants [1, 4, 5]. In addition, commercial mouthwashes are used as antiseptics for better oral hygiene. Therapeutic mouthwashes reduce bacterial counts, have antiplaque effects, work as an astringent, and help to reduce gingivitis and carious lesions [6, 7].

Dental materials should not be toxic, irritant, or corrosive and should be easy to use. The biomaterials used in dentistry can be metals (amalgam of silver, titanium, and gold), ceramics (feldspar, alumina, zirconia, porcelain reinforced with silica), composites [1], materials that can optimize dentin bonding, and mouthwashes. However, in the field of health, the use of natural products for dental biomaterials and curing of diseases has always emphasized, rather than depending on the conventional allopathic medicine [7].

Following this trend, in addition to the devices and materials themselves, biological advances have revolutionized the methods used in the chemical and material industries to produce and transform raw materials. Living plants can be processed in large quantities to produce a much larger variety of liquids and materials, without the cost of energy or effluent streams—typical by-products of the chemical industry. Nature is not only capable of allowing the synthesis of new chemical substances but also significantly reducing the costs and environmental impacts associated with the manufacture of current chemicals and drugs [3].

In 1978, the World Health Organization (WHO) recognized medicinal plants as a therapeutic resource [8]. The Ordinance No. 971 dated May 03, 2006, approved the National Policy on Integrative and Complementary Practices (PNPIC) in the Unified Health System (SUS) in Brazil, including the use of phytotherapy [9]. At present, phytotherapy is defined as a science-based practice for the treatment of diseases, which uses medicinal plants, plant drugs, and preparations, not including substances from another source [10]. Therefore, biomaterials made today are routinely information rich and incorporate biologically active components derived from nature [2, 3, 5, 7]. Today, the variety of natural products used in the biomaterials for dental and oral health care may include natural silk [11], propolis [12–15], chitosan [16], herbal tea [17], and miswak [18], as well as natural products for bone repair such as dolomite [4].

Since phytotherapy is a feasible method for the control and prevention of the development of oral pathologies, with the additional possibility of incorporating phyto-derived compounds into biomaterials, the discovery of new phytotherapic compounds has been of high relevance to dentistry [5, 14, 19–22]. Brazil has an advantage in this market because it has the greatest biodiversity in the world, especially in the Amazo nian Region, and a genetic heritage of great potential for the development of new herbal products [23], especially in dentistry. Given the growth of products derived from medicinal plants in Brazil, it was necessary to implement a statute that covered the requirements for all medicines and/or biomaterials to ensure the quality, efficacy, and safety of these products. In this sense, the Brazilian National Health Surveillance Agency (ANVISA) establishes product quality control requirements, involving stages ranging from obtaining of raw materials through to the qualitative and quantitative characterization of their active principles [24, 25].

The official recognition of phytotherapy in dentistry in Brazil was accompanied by several gaps in scientific research on medicinal plants, specifically for plant species with applications in diseases of the oral cavity. The state of Amazonas, specifically the city of Manaus, did not have a diagnosis of the applicability of medicinal plants in dental services. In this sense, these researchers conducted an ethnobotanical study to identify the main plants used for pathogenesis of the oral cavity, with the aim of reducing their empirical use and favoring the use of medicinal plants based on scientific evidence [10].

The search for the biomaterials, their development, and pharmaceutical forms comprises products derived from medicinal plants with compounds that are safe and have proven quality. Thus, researches in dentistry have been developed with the objective of searching for new bioactive principles for the formulation of drugs with different types of applications, capable of acting in both preventive strategies and curative treatments, thus encouraging the use of phytotherapic agents such as *Copaifera multijuga*, *Apis mellifera* (propolis), and *Libidibia ferrea*.

2. *Copaifera multijuga*—copaiba oil

According to the growing interest in antimicrobial agents derived from medicinal plants, natural products are considered an excellent alternative to synthetic chemicals. Amazonian biodiversity products that have been used for years in folk medicine have emerged as feasible and promising alternatives for inhibiting microorganisms in dental biofilm. Copaiba oil—as it is popularly called—a phytotherapeutic agent widely used by the Amazonian population, is known for its antibacterial, anti-inflammatory, anesthetic healing and antitumoral medicinal properties.

The studies developed with copaiba oil have complied with all the norms required. The *Copaifera multijuga* Hayne species were collected from official research institutions (National Research Institute of Amazonian—INPA) to guarantee the legitimacy of the species. The exsiccata was stored in the INPA Herbarium under No. 270709.

2.1 Antibacterial activity of copaiba oil formulations

The first reports demonstrated the use of copaiba oil as an effective agent against the etiological agents of caries disease, as seen in Ref. [26]. This research demonstrated the antibacterial activity of calcium hydroxide and zinc oxide pastes associated with essential oil and *C. multijuga* resin against *Streptococcus mutans*. The bacteriostatic and bactericidal activities of the oil *in natura* against the same microorganism were also reported, giving rise to a line of research in dentistry in the search of scientific evidence.

C. multijuga Hayne has presented promising antibacterial activity against *S. mutans, S. mitis*, *S. salivarius*, *S. constellatus*, and *S. sanguinis* from the copaiba gel production for the dental biofilm control [27]. In addition, evidence was shown of antibacterial activity against *Enterococcus faecalis* and *Candida albicans* present in endodontic flora, biological compatibility test in the teeth of rats and dogs using the copaiba oil as vehicle for calcium hydroxide [26], biological compatibility test in gingival tissue and a physical-chemical study and antibacterial activity dental cement to suit the conditions of the oral environment [28], and antibacterial activity in copaiba oil emulsions, as shown in **Table 1** [22].

The results of the chromatographic analysis of *C. multijuga* oil revealed that the structures of its components are made up of various sesquiterpenes, primarily being constituted of β-caryophyllene and its oxide, forming its biological activity [26–28].

Tested copaiba emulsions	MIC of the copaiba emulsions (μL/mL) against the bacteria			
	S. mutans (μL/mL)	*S. oralis* (μL/mL)	*S. salivarius* (μL/mL)	*L. casei* (μL/mL)
Emulsion 10%	12.5	12.5	12.5	12.5
Emulsion 10% + PB 1%				
Emulsion 30%	37.5	37.5	37.5	37.5
Emulsion 30% + PB 1%				

Table 1.
Minimum inhibitory concentration (MIC) of copaiba emulsions against microorganisms.

In [29], a high proportion of sesquiterpenes (88.55–98.05%) was revealed in the copaiba oils analyzed, with β-caryophyllene being the main type.

2.2 Copaiba oil emulsion as dentin biomodifier

Different approaches have been proposed to improve the restorative material bond to the dental structure by optimizing the infiltration of resinous monomers into the demineralized dentin and reduce the rate of water absorption and collagen degradation, by means of such as the application of an additional layer or multiple layers of a hydrophobic adhesive agent [30], vigorous solvent evaporation [31], polymerization, and the use of electric current to improve impregnation of the monomers [32].

Metalloproteinases (MMPs) trapped in the extracellular matrix are calcium-dependent and zinc-activated enzymes that mediate the denaturation of the extracellular matrix through collagenase (MMP-8 is the major collagenase in human dentin) and gelatinase (MMP-2 and MMP-9), as well as the enamelysin MMP-20 and the stromelysin MMP-3, which are naturally entrapped in the mineralized dentin during odontogenesis [33]. Since the bonding process occurs as a result of encapsulation of the collagen by the adhesive system, it is necessary to inhibit these enzymes to preserve the adhesive interface from proteolytic and hydrolytic degradation, by forming the hybrid layer [34].

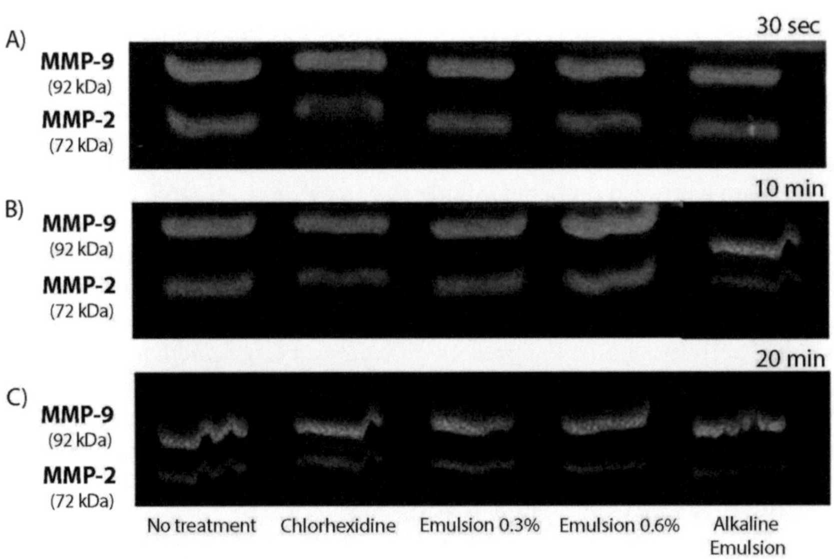

Figure 1.
Effect of copaiba oil on metalloproteinase (MMP) activity (gelatin zymography).

In recent studies, a zymography assay was performed with HT1080 cells. This demonstrated that the copaiba oil emulsion (CO) as dentin biomodifier showed the potential to inhibit matrix metalloproteinases −2 and −9 (**Figure 1**).

In the time of 30 seconds, statistical difference in the decrease of the MMP-9 activity was observed among all solutions tested when compared with the control group without treatment. In the time of 10 minutes, there was a statistical difference only between chlorhexidine (CLX), 10% CO + 0.3% CV, and 10% alkaline CO. At 20 minutes, the only solution that presented a statistical difference in the decrease of MMP-9 activity was the 10% CO + 0.3% CV. The best result obtained was with 10% alkaline CO in a time of 10 minutes, with a 35% decrease in MMP-9 activity (**Figure 2A**).

A decrease in the MMP-2 enzymatic activity was also observed. At 30 seconds, there was a statistical difference between the CLX solution and the copaiba oil emulsion at 0, when compared to the control group. At 10 minutes, a statistical difference was observed in the CLX solution and the alkaline CO, each presenting a 44 and 67% reduction in enzyme activity, respectively. Finally, at 20 minutes, there was a statistical difference in the CLX solution (21%), in the COs 0.3% (42%) and 0.6% (53%), and in the alkaline CO (**Figure 2B**).

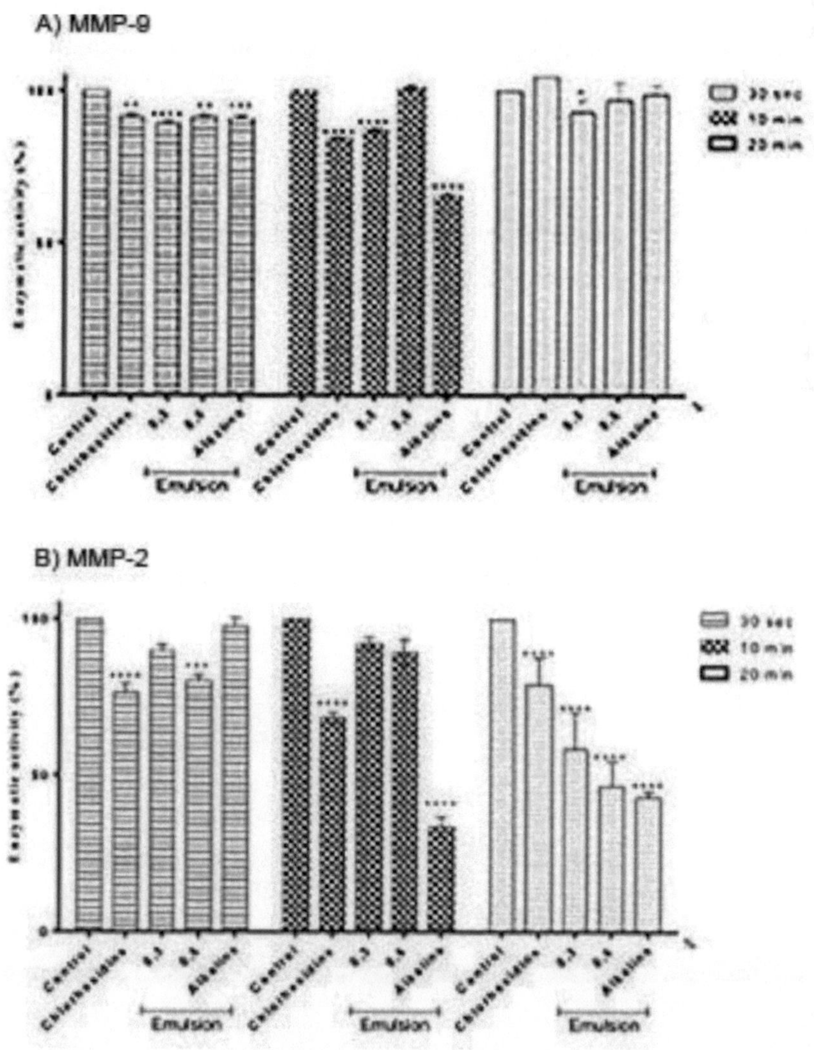

Figure 2.
Anti-proteolytic activity of the copaiba oil emulsions on (A) MMP-9 and (B) MMP-2.

2.3 Influence of copaiba oil emulsions as dentin surface biomodifiers

Bandeira et al. [5] used scanning electron microscopy (SEM) to investigate the morphology of the dentin surface, cut and treated with CO and ethanolic extract of propolis, with the aim of using them as bioactive agents for cleaning teeth. For the formulations of 10% CO, 10% CO + PB, 30% CO, and 30% CO + PA emulsions, the same cleaning pattern was obtained as that obtained with 2% chlorhexidine, which is considered the gold standard, because it had substantivity and showed bacteriostatic and bactericidal activities (**Figure 3**).

The bond of polymer-based materials to dentin is still considered a significant challenge because the latter is a complex substrate, predominantly tubular, and intrinsically moist. The use of disinfectant and anti-proteolytic solutions may be an alternative for reducing these effects. In a histological evaluation, 10% CO was used on the exposed collagen of the dentin matrix, with the purpose of verifying whether there was interference in the adhesive system. Thus, 80 specimens (CPs) were prepared from healthy third molars, and after the induction of caries lesions, the specimens were treated with test materials for 3 months of aging. The CPs treated with the copaiba emulsions presented higher exposed and hybridized collagen thickness values than the groups treated with CLX 2% and AD. Relative to caries-affected dentin, the group treated with CLX 2% showed a higher proportion of CPs with hybridized collagen. The emulsion presented 100% specimens with hybridized collagen and improved hybrid layer homogeneity (**Figure 4**).

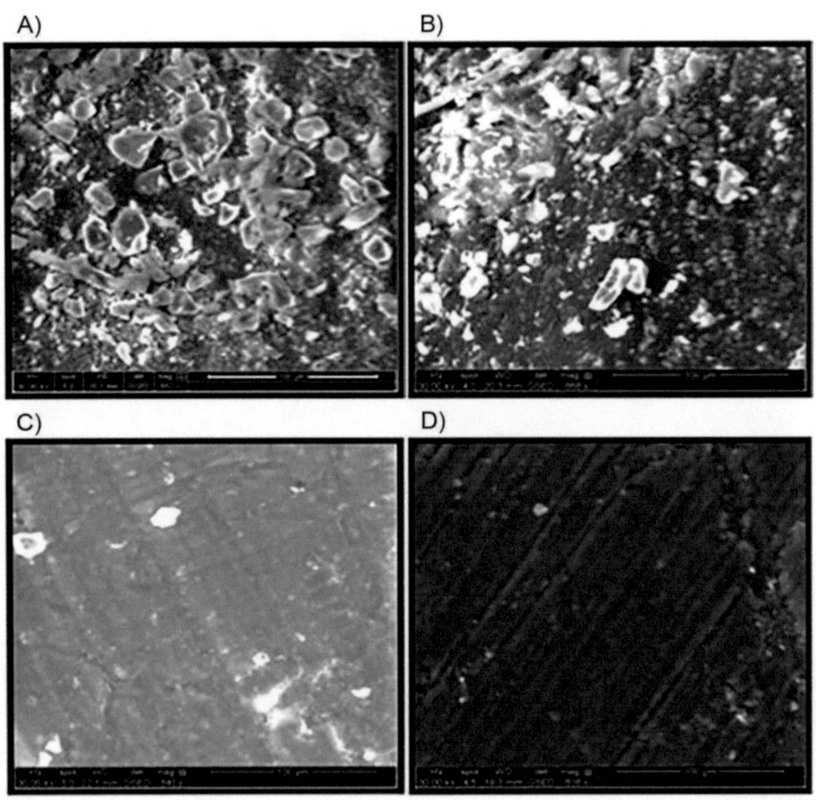

Figure 3.
Photomicrograph of the group (A) without dentin surface treatment, (B) air/water spray treatment, (C) chlorhexidine, and (D) copaiba oil emulsion treatments.

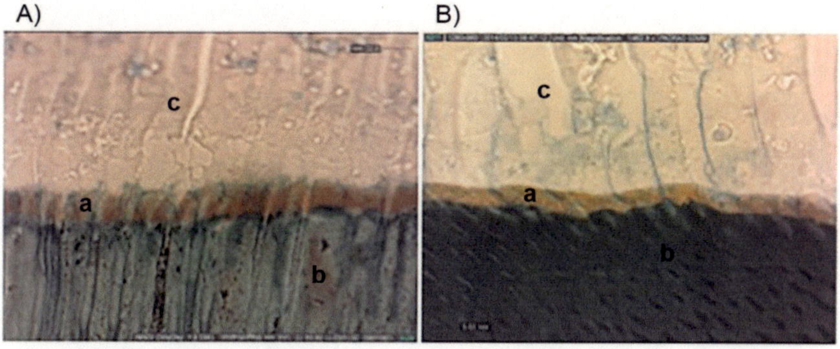

Figure 4.
Hybrid dentin specimen stained with Goldner's Trichrome. (A) Chlorhexidine treatment and (B) copaiba oil emulsion treatment; (a) hybridized collagen, (b) mineralized dentin, (c) adhesive.

Given the biological properties of *C. multijuga* Hayne, a copaiba oil emulsion was formulated for use before applying the adhesive to improve the quality of dentin bonding. The morphological characteristics of the dentin surface and the hybrid layer formed with etch-and-rinse and self-etching adhesive systems in healthy and caries-affected dentin after using the 10% copaiba oil emulsion were analyzed. A total of 96 third molars from Biobank of the School of Dentistry, Federal University of Amazonas (FAO-UFAM), were used. Half of the teeth underwent artificial induction of dental caries and the other half formed the group of healthy teeth. The roots of all the teeth were removed, yielding dentine disks that were divided into groups according to test substances (CLX 2%, copaiba oil emulsion, calcium hydroxide solution, and distilled water), the dentin (sound or caries-affected), and the adhesive system (Adper Single Bond 2® and Clearfil SE Bond®).

SEM was used to analyze the dentin surface and hybrid layer of the specimens obtained, according to the experimental groups. The dentin surface treatment with copaiba oil emulsion showed no physical barrier to adhesive penetration. The dentin surface treated with 2% chlorhexidine showed phosphate salts in two types of dentin. Dentin surface treatment with calcium hydroxide solution resulted in the deposition of mineral precipitate obstructing the lumen of the tubules in sound dentin. The result of calcium hydroxide solution applied on the conditioned sound dentin differed from those of the other substances ($p < 0.05$). On the smear layer surface, the result of distilled water on sound dentin showed a significant difference from the results of all experimental groups ($p < 0.05$). There was no statistical difference between the hybrid layer formed with the Single Bond® adhesive in otherwise healthy dentin specimens and those of caries-affected dentin treated with the test substances.

In the dentin-rich and caries-affected dentin treated with CLX 2%, the hybrid layer formed with Adper Single Bond® was thick, regular, and homogeneous, with long resin tags, but in smaller quantity than those in the distilled water group. The hybrid layer formed with Clearfil SE Bond® on the dentin treated with CLX showed few irregular resin tags with little adhesive infiltration into the dentin (**Figure 5**).

The hybrid layer formed with the Clearfil SE Bond® adhesive on the caries-affected dentin and caries-affected dentin treated with the copaiba oil-based emulsion presented regular and homogeneous hybrid layer with a large number of resin tags (**Figure 6**).

The CO application showed no morphological change in sound and caries-affected dentin, irrespective of phosphoric acid etching, and presented a uniform hybrid layer, regular, and extensive monomer infiltration into sound and caries-affected dentin, irrespective of the adhesive system.

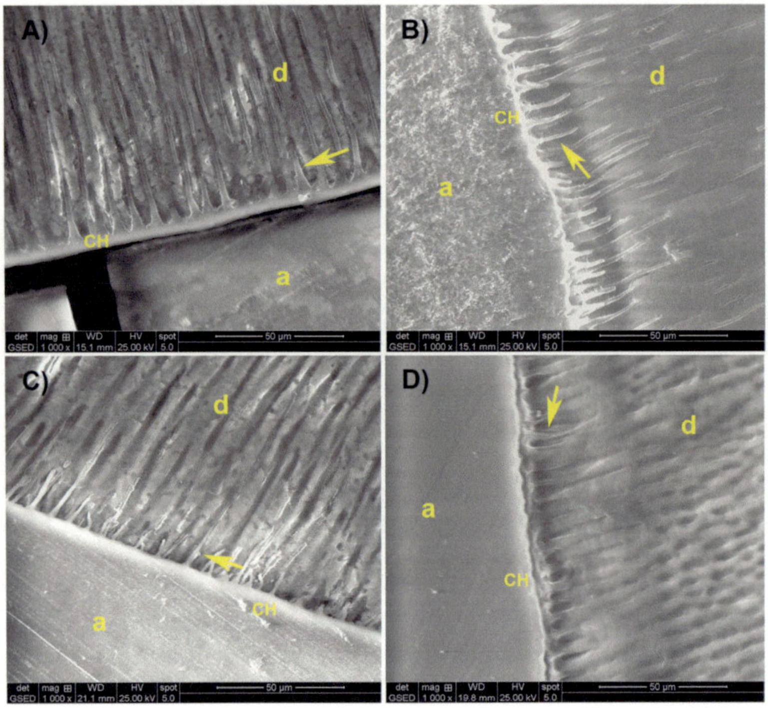

Figure 5.
Hybrid layer formed with Single Bond in the dentin treated with chlorhexidine: (A) Healthy dentin and (B) caries-affected dentin. Hybrid layer formed with Clearfil on the dentin treated with chlorhexidine: (C) Healthy dentin and (D) caries-affected dentin. (a) Adhesive, (d) Dentin, (CH) Hybrid layer. The arrows point to the resin tags.

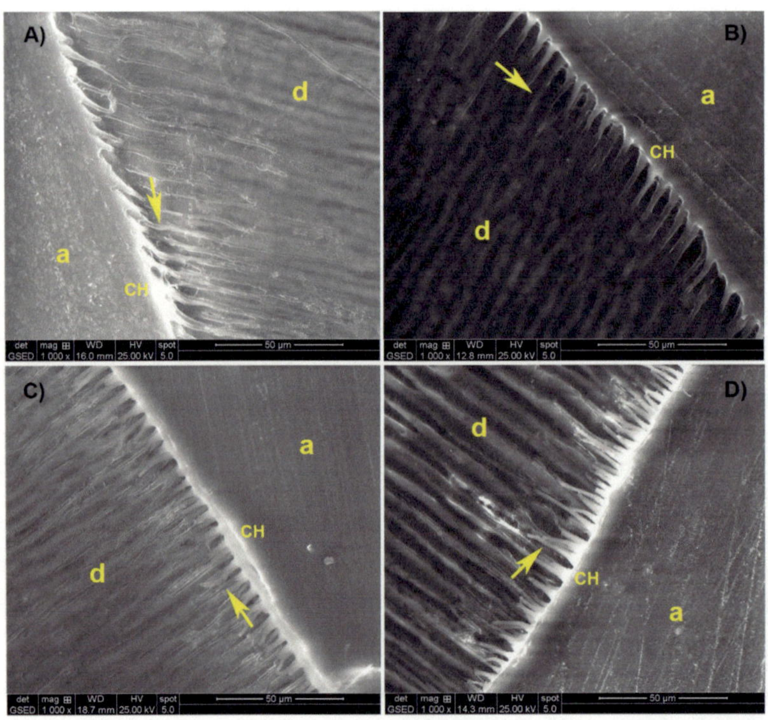

Figure 6.
Hybrid layer formed with Single Bond in dentin treated with copaiba oil: (A) healthy dentin and (B) caries-affected dentin. Hybrid layer formed with Clearfil in dentin treated with copaiba oil: (C) healthy dentin and (B) caries-affected dentin. (a) adhesive, (d) dentin, and (CH) hybrid layer. The arrows point to the resin tags.

The hybrid layer formed with Adper Single Bond® in the carious-affected and caries-affected dentin treated with copaiba oil-based emulsion was thick, regular, and homogeneous, with long resin tags, but in smaller quantity than those in the distilled water group (**Figure 6**).

Thus, the copaiba oil emulsion as dentin biomodifiers with their antibacterial activity and property of inhibiting MMPs may contribute to stability of the hybrid layer, perhaps because they prevent the enzymatic hydrolysis of collagen due to their oily nature similar to that of mineral oil. Further studies must be conducted to show the mechanism of action of the oil on the MMPs and the formation of the hybrid layer. In the present work, it was shown that the MMPs were resistant to the time-dependent destruction of the hybrid layer and that the use of inhibitors could improve the durability of the composite resin-dentin bond [35].

3. *Apis mellifera*—Propolis

Propolis is a resinous substance, collected and transformed by the *Apis mellifera* bees, used in the hives as sealant. The vegetation, climate, and other factors influence the characteristics and composition of the propolis of each region. Studies have shown that propolis is composed of more than 300 (three hundred) substances, among them the most important are flavonoids and their phenolic compounds. Its biological properties include anticancer, antioxidant, anti-inflammatory, antibiotic, and antifungal activities [12, 13].

Worker bees of the *A. mellifera* species remove propolis from the shoots and resinous secretions present in the trees. By using their jaws and their legs, they mix these secretions with the wax synthesized by them. The mandibular glands of the workers secrete 10-hydroxydecanoic acid, which enables the finishing of propolis. Factors such as plant ecology of the region where the propolis was collected and even the genetic variability of the queens also influence the chemical composition of propolis. According to CONAPIS (information body of CONAP, National Council of Apiculture Ltda.), the ideal propolis is the type produced in regions where there is as little environmental pollution as possible, away from urban centers and factories that emit pollutants [14].

Almeida et al. [15] reported that propolis is a complex mixture of resinous, gummy, and balmy substances collected by *A. mellifera* bees. It has a bactericide, antimicrobial, antioxidant, anti-inflammatory, immunomodulatory, hypotensive, anesthetic, anti-cancerous, anti-HIV, and anti-carious biologic function. These biological activities are related to their chemical components present in the propolis of the Amazon region.

Ishida et al. [14] analyzed the ethanolic propolis extracts from four propolis samples (E1–E4) from Manaus (Brazilian Amazon) by HPLC/DAD/ESI–MS/MS and GC/EIMS. The main components of E2 and E4 were polyprenylated benzophenones: 7-epi-nemorosone, 7-epi-clusianone (major E4 constituents), xanthochymol, and gambogenone (major E2 constituents), making up a chemical profile so far unreported for Brazilian propolis. Aristhophenone, methyl insignimone, 18-ethyloxy-17-hydroxy-17, 18-dihydroscrobiculatone B, and derivatives of dimethyl weddellianone A and B, propolones, and a scrobiculatone derivative were detected as minor constituents. Triterpenoids (b-amyrins, b-amyrenone, lupeol, and lupenone) were ubiquitous and predominant in E1 and E3. The extracts E2 and E4 were highly active against the cariogenic bacteria *S. mitis*, *S. mutans*, and *S. salivarius*. E2 was more active than E4, probably due to a higher content of 2-epi-nemorosone, while the latter was more abundant in dihydroxylated compounds.

By histological analysis of the extracts in subcutaneous connective tissue in rats, a propolis solution for cavity cleansing and its toxicity was investigated through hemolytic and *Artemia franciscana* tests. Fifteen male rats were selected and randomly distributed into three experimental time intervals (07, 30, and 45 days), in which each animal received the four groups of treatment in rounds: Group I, Propolis I; Group II, Propolis II; Group III, calcium hydroxide water; and Group IV, 2% CLX; the sides of the tube were the control group. As regards biocompatibility, the results showed that all materials presented a significant reduction in inflammatory infiltrate and an increase in collagen fiber thickness values (**Figure** 7). In decreasing order of biocompatibility, the use of the following materials may be suggested: calcium hydroxide-water, 2% chlorhexidine®, Propolis I, and Propolis II. In the cytotoxicity test using *Artemia franciscana*, the propolis extract showed high toxicity in the tested concentrations, and in the hemolytic activity test, the Propolis I extract showed more activity than Propolis II. Therefore, the present study suggested the use of propolis as a cavity cleansing solution for shallow and medium cavities similarly to the use of 2% chlorhexidine® [15].

In continuing studies of the application of biomaterials in dentistry, the response of inflammatory periodontal disease (PD) induced in rat periodontal tissue was histologically evaluated after the use of 0.1, 1, and 10% aqueous suspensions of propolis (SAP) for subgingival irrigation. A total of 84 Wistar rats (*Rattus norvegicus*) were distributed into the following experimental groups: Group I (n = 14, 0.1% SAP), Group II (n = 14, 1% SAP), Group III (n = 14, 10% SAP), Group IV (n = 14, 5% Tween 80 solution), Group V (positive control, n = 14, with induced and untreated PD), Group VII (n = 14, 2% CLX), and Group VI (n = 84, negative control, non-induced and unprocessed contralateral teeth of Groups I, II, III, IV, V, and VII). After induction of PD by the ligation technique in the cervical portion of the mandibular left first molar for 15 days, the periodontal pocket was irrigated three times (first, fourth, and seventh days). The animals were sacrificed at 15 and 30 days after the treatment. The results suggested that SAP demonstrated an effective inflammatory response in the short term (15 days) and in the concentration of 0.1% was associated with the presence of dense gingival fibers, blood vessels

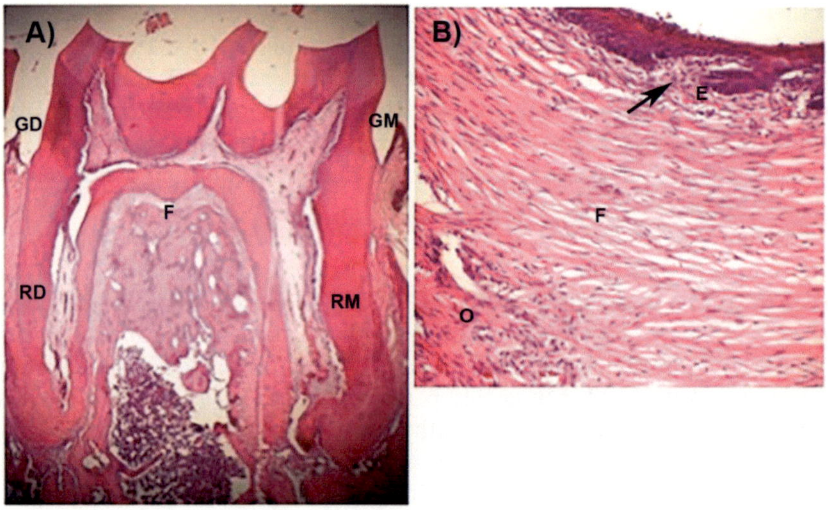

Figure 7.
(A) Anatomical shape of rat molar (H.E. 40x): (RM) mesial root, (RD) distal root, (F) furca, (GM) mesial gingiva, and (GD) interproximal gingiva. (B) Furcation region of the tooth treated with aqueous suspension of 0.1% propolis at 30 days (H.E. 200x): (E) epithelium of the exocytosis purse-string, (F) moderate collagen fibers, (O) bone tissue, and (arrow) moderate, chronic, and focal inflammatory infiltrate.

without congestion, absence of dental resorption and bone loss, and may be an alternative treatment of PD. However, future studies are required to demonstrate the biological feasibility of the use of propolis as an adjuvant to periodontal therapy.

4. *Libidibia ferrea*

4.1 Plant species and its applicability

Among the vast biodiversity of medicinal plants, *Libidibia ferrea*—popularly referred to as Jucá or ironwood—is another plant from the Brazilian biome noted for several therapeutic properties [20]. Characterized as belonging to the Leguminosae family tree, its scientific name is *Libidibia ferrea* (Mart. ex Tul.) L.P.Queiroz *var. ferrea*, ex *Caesalpinia ferrea* Mart. ex Tul. (International Plant Names Index, 2009) [36].

The fruits (pods) of this species are used for the treatment of diabetes and cancer prevention, in addition to wound healing [37]; the roots have been documented as having antipyretic effects, being used in the treatment of diarrhea, and having anticancer properties [38, 39]; and the bark has been used for treatment of enterocolitis and rheumatism [40]. Therefore, these parts of this species have shown anti-inflammatory [36], antifungal [19], antihistaminic, antiallergic, anticoagulant [41], antiproliferative, cytoprotective, and antimutagenic effects [42]. Furthermore, *L. ferrea* has been used for biosynthesis of silver nanoparticles (AgNPs), thereby preserving its antimicrobial activity, reducing its toxic effects on human cell lines and increasing its practical use without the impact on the environment. AgNPs are one of the most important nanomaterials among several metallic nanoparticles that are involved in biomedical applications [23].

In dentistry, *L. ferrea* has demonstrated potential antibacterial activity against oral microorganisms [20, 21], in addition to anti-inflammatory and analgesic properties [41, 43]. Nevertheless, several tests are still necessary to improve new dental products and prove their efficacy when used in the oral cavity, with the aim of using them in the dental clinic.

4.2 Preliminary studies—standardization of the extract

Initially, taking into consideration the vast biodiversity of medicinal plants of the Amazon region (Brazil), which have been used empirically due to their antibacterial action, these species have been screened with regard to their antimicrobial activity against microorganisms isolated from dental biofilm [44].

Since the research began, there has been constant concern to ensure the legitimacy of plant species used to obtain the study extracts. The species were collected from official research institutions (Brazilian Agricultural Research Corporation, EMBRAPA; Federal University of Amazonas, UFAM; and Brazilian Institute of Environment and Renewable Natural Resources, IBAMA) and were stored at the Lauro Pires Xavier Herbarium in the Systematic and Ecology Department/Federal University of Paraíba, according to the Genetic Heritage Component Sample Access and Delivery Authorization (No. 044/2004–IBAMA/MMA). From this initial screening, both *L. ferrea* and *C. multijuga* extracts exhibited antimicrobial activity against *S. sobrinus, S. mutans, S. mitis, S. sanguis*, and *Lactobacillus casei* strains and inhibited microbial adherence of these tested strains (**Table 2**) [44].

Sequentially, studies have shown *L. ferrea* fruit extract inhibited the in vitro growth of the following oral pathogens (*C. albicans*, *S. mutans, S. salivarius, S. oralis*, and *L. casei*) on planktonic cells and multispecies biofilm models, supporting the use of this extract for the treatment of oral infections [19]. On the other

	Microorganisms				
Tested substances/MIC	***S. mutans***	***S. mitis***	***S. sanguis***	***S. sobrinus***	***L. casei***
L. ferrea (Jucá) 1:8 (62.5 mg/mL)	15	14	14	15	15
A. chica (Crajiru) (500 mg/mL)	0	0	0	13	13
O. micranthum (alfavaca) (1000 mg/mL)	0	0	16	12	0
C. multijuga (copaíba) (1000 mg/mL)	20	19	19	20	22
Chlorhexidine 0.12% 1:8 (0.15 mg/mL)	12	13	13	12	13
A. nitidum (carapanaúba)	0	0	0	0	0
C. guianensis (andiroba)	0	0	0	0	0
E. ayapana (Japana-branca)	0	0	0	0	0
S. acmella (jambu)	0	0	0	0	0

Table 2.
Diameter of the inhibition halos (mm) of the substances against the microorganisms.

hand, *L. ferrea* stem bark extract exhibited better antimicrobial activity than fruit extract when tested against the same oral microorganisms in planktonic cells [45].

Dental biofilm is a dense, whitish, noncalcified aggregate of bacteria, with desquamated epithelial cells and food debris creating conditions for an imbalance of resident oral microflora, favoring the destruction of hard and soft tissues by the development of oral pathologies such as caries and gingivitis. Recently, an *L. ferrea* extract was standardized according to the current Brazilian legislation with regard to pH, sedimentation, density, and stability, along with microbiological tests of the extract. The microbial test was used to verify the presence of *Staphylococcus aureus*, *Pseudomonas aeruginosa*, fungi, yeasts, coliforms, and minimum inhibitory concentrations of *S. mutans* and *S. oralis* strains. Thus, this *L. ferrea* extract was shown to have antibacterial activity against the oral microorganisms tested and satisfactory stability and quality, enabling the formulation of a mouthwash using this extract to control dental biofilm [20].

4.3 Standardization of the orabase and mouthwash formulations

After the preliminary studies of the *L. ferrea* extract, orabase and mouthwashes based on this extract were analyzed.

The use of oral antimicrobial formulations as an adjunct treatment to mechanical means of dental biofilm and gingival inflammation control has been well established [46]. However, acceptance of the use of plant-based oral products still faces obstacles due to a lack of quality control, since the profile of the end-product constituents has implications in phytotherapeutic efficiency and safety [47]. Thus, the Brazilian ANVISA [48] has established that all phytotherapeutic medication must be submitted to formulation stability tests. Production operations must follow operational procedures with clearly defined and approved standards, in conformity with the notification or registration of traditional phytotherapeutic products with the competent sanitary agency. The final objective is to obtain products that are within the quality standards demanded.

Therefore, Marreiro et al. [20], in a preliminary study, evaluated the antimicrobial activity of aqueous extracts of the fruits, stem bark, and an orabase formulation of *L. ferrea* against biofilm microorganisms by the agar diffusion and broth microdilution methods and evaluated cytotoxicity by hemolysis assay on fibroblast cell culture. This study endeavored to find an alternative material as a way to guarantee the supply of raw vegetable matter independent of the seasonality of the fruits. The microorganisms used for determining the MICs were *S. salivarius*, *S. mutans*, *S. oralis*, *L. casei,* and *C. albicans*. The extract of the stem bark and the orabase formulation of *L. ferrea* showed antimicrobial activity against microorganisms of the biofilm and were not toxic when tested on erythrocytes and in cell cultures (**Figure 8**).

In addition to these results, Venâncio et al. [21] evaluated the in vitro pharmacological stability of a phytotherapeutic mouthwash based on *L. ferrea* extract with regard to the microbiological parameters of control, organoleptic characteristics, sedimentation, pH, and density. Using methods in accordance with the legislation, the study determined the total number of microorganisms and *Salmonella* sp., *Escherichia coli*, *Pseudomonas aeruginosa*, and *Staphylococcus aureus*; stability characteristics (color, odor, brightness, and consistency); sedimentation test (centrifuge); the pH measurement (pH meter); and density evaluation (pycnometer). The results demonstrated the *L. ferrea* mouthwash was shown to be free of contamination for the tested microorganisms and was within the standards of safety demanded for its use.

In addition to the reported studies, with the aim of finding another purpose for its use, Matos [49] evaluated a formulation of *L. ferrea* orabase for use in the healing of oral ulcers. The overall aim of this study was the physical-chemical quality control of an orabase ointment formulation of *L. ferrea*. For physicochemical evaluation, centrifugal tests, pH, mass, relative density, microbiological assessment, and organoleptic character of contaminant tests were performed. The physical conditions tested were storage at room temperature (±25.9°C), room temperature, protected from light (±28.8°C), and air conditioning (±23.7°C); the experimental time intervals were 0, 30, and 60 days. The results showed that in the centrifugation test, phase separation was observed at all times and in all storage environments. In the pH test, only the formulation stored at room temperature obtained a lower pH value, mean (1.95) after 60 days of manipulation. The density test showed the mean value of 0.809 g/cm^3 when tested at time 0 and after 30 days of formulation values under air conditioning (0.746 g/cm^3), room temperature (0.702 g/cm^3), and room temperature, protected from light, dark (1.022 g/cm^3). The evaluation of contaminants showed that there was no bacterial growth in any environment and experimental time, but macroscopically, the increase in cotton wool colonies compatible with fungal colonies was observed in the formulation stored under air conditioning in the time interval of 30 days and under all storage locations in the time interval of 60 days. In the organoleptic assessment, the ointment showed changes at 60 days after formulation. Based on the results, the formulation tested maintained the best stability of the tested characteristics when stored at room temperature in the dark. However, after 60 days of storage, the formulation presented chemical and physical instability and growth of contaminants.

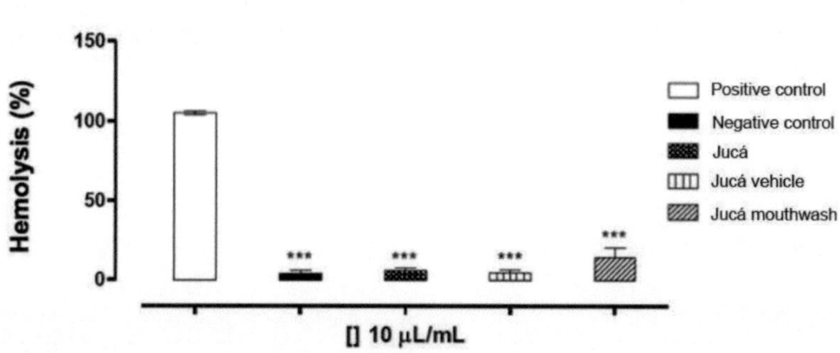

Figure 8.
*Hemolysis assay: Triton™ X-100 (positive control). *** Statistically significant in comparison with the positive control (P < 0.05).*

5. Conclusions

Given the growth of the products derived from medicinal plants in Brazil, it was necessary to implement a statute that covered the requirements for all medicines and biomaterials to ensure the quality, efficacy, and safety of these products. Based on the researches, *C. multijuga,* in the form of copaiba oil, was an effective agent against the etiological agents of caries disease. Copaiba oil emulsions as dentin biomodifiers with their properties of antibacterial activity and inhibition of MMPs may contribute to the stability of the hybrid layer in caries-affected dentin treated with this emulsion.

Propolis demonstrated an effective and controlled inflammatory response in the periodontal tissue such as the absence of bone resorption, blood vessels without congestion, and presence of dense gingival fibers.

The *L. ferrea* extract showed antibacterial activity against the oral microorganisms tested, and satisfactory stability and quality, enabling the formulation of a mouthwash using this extract to control dental biofilm. The results demonstrated that the *L. ferrea* mouthwash was shown to be free of contamination for the tested microorganisms and was within the standards of safety demanded for/to allow its use.

As propolis showed a reduction in bone resorption, and the other medicinal plants studied were in compliance with the safety standards, future investigations into the effectiveness of adding these herbal medicines to bone substitutes will be conducted. Via an effective biodegradation delivery system, the release of phytotherapic agents is expected to promote the growth of bone tissue. Pharmacological tests will examine the phytotherapic ability to generate an osteotropic effect, stimulating bone cell proliferation and differentiation while decreasing osteoclast activity. Thus, future studies will be in search of herbal devices that stimulate the innate regenerative capacity of bone and can be used in the regeneration of bone tissue.

Acknowledgements

This work was supported by grants from the National Science Centre (NCN)—SCN (UMO-2013/10/E/NZ3/007500) and SYMFONIA 3 (UMO-2015/16/W/NZ4/00071)—and the National Centre for Research and Development (NCBR)—STRATEGMED III (BioMiStem 303570) to EZS. The Faculty of Biochemistry, Biophysics, and Biotechnology at the University, Krakow, Poland, is a partner of the Leading National Research Center supported by the Ministry of Science and Higher Education and National Council for Scientific and Technological Development (Processes No 575752/2008-4, No. 575676/2008-6 and CT Amazon 2013 No. 406457/2013-1).

Conflict of interest

The authors declare no competing interests.

Author details

Maria Fulgência Bandeira[1*], Nikeila Conde[1], Carina Toda[1], Ana Regina Casaroto[1], Geisy Lima[1], Marne Vasconcellos[2], Leilane Bentes[2], Larissa Silva[1], Samuel Lungareze[1] and Joyce Meira[1]

1 School of Dentistry, Federal University of Amazonas, Manaus, Brazil

2 School of Pharmaceutical Sciences, Federal University of Amazonas, Manaus, Brazil

*Address all correspondence to: fulgencia@ufam.edu.br

References

[1] Clinical Applications of Biomaterials. National Institute of Health Consensus Development Conference Statement [Internet]. 1982. Available from: https://consensus.nih.gov/1982/1982Biomaterials034html.htm [Accessed: 25-07-2018]

[2] Anderson JM, Rodriguez A, Chang DT. Mint: Foreign body reaction to biomaterials. Seminars in Immunology. 2008;**20**:86-100. DOI: 10.1016/j.smim.2007.11.004

[3] Huebsch N, Mooney DJ. Mint: Inspiration and application in the evolution of biomaterials. Nature. 2009;**462**:426-432. DOI: 10.1038/nature08601

[4] Moreschi E, Hernandes L, Dantas JA, da Silva MA, Casaroto AR, Bersani-Amado CA. Mint: Effect of dolomite on the repair of bone defects in rats: Histological study. Histology and Histopathology. 2010;**25**:1547-1556. DOI: 10.14670/HH-25.1547

[5] Bandeira MFCL, Lima GR, Lopes PP, Toda C, Venâncio GN, Lima GA, et al. Mint: Dentin cleaning ability of an amazonian bioactive: Evaluation by scanning electron microscopy. The Open Dentistry Journal. 2016;**10**(Suppl-1, M5):182-187. DOI: 10.2174/1874210601610010182

[6] Prasad KR, John S, Deepika V, Dwijendra KS, Reddy BR, Chincholi S. Mint: Anti-plaque efficacy of herbal and 0.2% chlorhexidine gluconate mouthwash: A comparative study. Journal of International Oral Health. 2015;7:98-102. PMCID: PMC4588801

[7] Khurshid Z, Naseem M, Zafar MS, Najeeb S, Zohaib S. Mint: Propolis: A natural biomaterial for dental and oral healthcare. Journal of Dental Research, Dental Clinics, Dental Prospects. 2017;**11**:265-274. DOI: 10.15171/joddd.2017.046

[8] WHO Traditional Medicine Strategy 2014-2023. World Health Organization [Internet]. 2013. Available from: http://www.searo.who.int/entity/health_situation_trends/who_trm_strategy_2014-2023.pdf?ua=1 [Accessed: 08-08-2018]

[9] Brazil. Ministry of Health. Ordinance no 971 of May 3, 2006. Available from: http://bvsms.saude.gov.br/bvs/saudelegis/gm/2006/prt0971_03_05_2006.html [Accessed: 20-08-2018]

[10] Evangelista SS, Sampaio FC, Parenter RC, Bandeira MFCL. Mint: Phytotherapics in odontology: Ethnobotanical study in Manaus. Revista Brasileira de Plantas Medicinais. 2013;**15**:513-519. DOI: 10.1590/S1516-05722013000400007

[11] Zafar M, Khurshid Z, Almas K. Mint: Oral tissue engineering progress and challenges. Journal of Tissue Engineering and Regenerative Medicine. 2015;**12**:387-397. DOI: 10.1007/s13770-015-0030-6

[12] Casaroto AR, Hidalgo MM, Sell AM, Franco SL, Cuman RK, Moreschi E, et al. Mint: Study of the effectiveness of propolis extract as a storage medium for avulsed teeth. Dental Traumatology. 2010;**26**:323-331. DOI: 10.1111/j.1600-9657.2010.00879.x

[13] Casaroto AR, Lara VS. Mint: Phytomedicines for Candida-associated denture stomatitis. Fitoterapia. 2010;**81**:323-328. DOI: 10.1016/j.fitote.2009.12.003

[14] Ishida VFC, Negri G, Salatino A, Bandeira MFCL. Mint: A new type of Brazilian propolis: Prenylated benzophenones in propolis from Amazon and effects against cariogenic bacteria. Food Chemistry. 2011;**125**:966-972. DOI: 10.1016/j.foodchem.2010.09.089

[15] Almeida MC, Ferreira LCL, Venâncio GN, Souza RS, Lima ES, Conde NC, et al. Mint: Biocompatibility of propolis in subcutaneous tissue of rats: A possible biomaterial for cavity cleansing. Revista Odonto Ciência. 2017;**32**:60-65. DOI: 10.15448/1980-6523.2017.2.26479

[16] Qasim SB, Najeeb S, Delaine-Smith R, Rawlinson A, Rehman IU. Mint: Potential of electrospun chitosan fibers as a surface layer in functionally graded GTR membrane for periodontal regeneration. Dental Materials. 2017;**33**:71-83. DOI: 10.1016/j.dental.2016.10.003

[17] Khurshid Z, Zafar MS, Zohaib S, Najeeb S, Naseem M. Mint: Green tea (Camellia sinensis): Chemistry and oral health. The Open Dentistry Journal. 2016;**10**:166-173. DOI: 10.2174/1874210601610010166

[18] Niazi F, Naseem M, Khurshid Z, Zafar MS, Almas K. Mint: Role of Salvadora persica chewing stick (miswak): A natural toothbrush for holistic oral health. European Journal of Dentistry. 2016;**10**:301-308. DOI: 10.4103/1305-7456.178297

[19] Sampaio FC, Pereira M do S, Dias CS, Costa VC, Conde NC, Buzalaf MA. Mint: In vitro antimicrobial activity of *Caesalpinia ferrea* Martius fruits against oral pathogens. Journal of Ethnopharmacology. 2009;**124**:289-294. DOI: j.jep.2009.04.034

[20] Marreiro RO, Bandeira MF, de Souza TP, de Almeida MC, Bendaham K, Venâncio GN, et al. Mint: Evaluation of the stability and antimicrobial activity of an ethanolic extract of *Libidibia ferrea*. Clinical, Cosmetic and Investigational Dentistry. 2014;**6**:9-13. DOI: 10.2147/CCIDE.S54319

[21] Venâncio GN, Rodrigues IC, Souza TP, Marreiro RO, Bandeira MFCL, Conde NCO. Mint: Herbal mouthwash based on *Libidibia ferrea*: Microbiological control, sensory characteristics, sedimentation, pH and density. Revista de Odontologia da UNESP. 2015;**44**:118-124. DOI: 10.1590/1807-2577.1064

[22] De Bari CCD, Sampaio F, Conde NCO, Moura L, Veiga Júnior V, Barbosa G, et al. Mint: Amazon emulsions as cavity cleansers: Antibacterial activity, cytotoxicity and changes in human tooth color. Revista Brasileira de Farmacognosia. 2016;**26**:497-501. DOI: 10.1016/j.bjp.2016.03.010

[23] Soares MRPS, Corrêa RO, Stroppa PHF, Marques FC, Andrade GFS, Corrêa CC, et al. Mint: Biosynthesis of silver nanoparticles using *Caesalpinia ferrea* (Tul.) Martius extract: Physicochemical characterization, antifungal activity and cytotoxicity. PeerJ. 2018;**19**:e4361. DOI: 10.7717/peerj.4361

[24] Brazil. Federal Council of Pharmacy. Ordinance no 138 of May 29, 2003. Available from: http://www.cff.org.br/userfiles/33%20-%20BRASIL_%20MINISTÉRIO%20DA%20SAÚDE%202003%20RDC_138_2003_ANVISA.pdf [Accessed: 20-08-2018]

[25] Brazil. National Health Surveillance Agency. Ordinance no 14 of March 31, 2010. Available from: http://www.crfma.org.br/site/arquivos/legislacao/resolucoeseinstrucoesnormativasdaanvisa/RDC%2014%202010.pdf [Accessed: 20-08-2018]

[26] Bandeira MF, Oliveira MR, Pizzolitto AC, Benatti Neto C. Mint: Preliminary study of the activity antibacterial of the essential oil and of the resin of the *Copaifera multijuaga* (copaiba oil), associated to the zinc oxide and to calcium hydroxide. Journal Brasileiro De Clinica & Estética Em Odontologia. 1999;**3**:46-51

[27] Simões CACG, Conde NCO, Venâncio GN, Milério PSLL, Bandeira MFCL, Veiga Júnior VF. Mint: Antibacterial activity of copaiba oil gel on dental biofilm. The Open Dentistry Journal. 2016;**10**:188-195. DOI: 10.2174/1874210601610010188

[28] Vasconcelos KR, Veiga Júnior VF, Rocha WC, Bandeira MF. Mint: In vitro assessment of antibacterial activity of a dental cement constituted of a *Copaifera multijuga* Hayne oil-resin. Revista Brasileira de Farmacognosia. 2008;**18**:33-38. DOI: 10.1590/S0102-695X2008000500017

[29] Souza BPC, Moreira WLS, Da Silva MR, De Tarso BSP, Vieira G, Veiga-Júnior VF. Mint: Phytochemical fingerprints of copaiba oils *(Copaifera multijuga* Hayne) determined by multivariate analysis. Chemistry & Biodiversity. 2013;**10**:1350-1360. DOI: 10.1002/cbdv.201200356

[30] Hashimoto M, Sano H, Yoshida E, Hori M, Kaga M, Oguchi H, et al. Mint: Effects of multiple adhesive coatings on dentin bonding. Operative Dentistry. 2004;**29**:416-423

[31] Van Landuyt KL, De Munck J, Snauwaert J, Coutinho E, Poitevin A, Yoshida Y, et al. Mint: Monomer-solvent phase separation in one-step self-etch adhesives. Journal of Dental Research. 2005;**84**:183-188. DOI: 10.1177/154405910508400214

[32] Pasquantonio G, Tay FR, Mazzoni A, Suppa P, Ruggeri A Jr, Falconi M, et al. Mint: Electric device improves bonds of simplified etch-and-rinse adhesives. Dental Materials. 2007;**23**:513-518. DOI: 10.1016/j.dental.2006.03.009

[33] Mazzoni A, Pashley DH, Tay FR, Gobbi P, Orsini G, Ruggeri A Jr, et al. Mint: Immunohistochemical identification of MMP-2 and MMP-9 in human dentin: Correlative FEI-SEM/TEM analysis. Journal of Biomedical Materials Research. Part A. 2009;**88**:697-703. DOI: 10.1002/jbm.a.31920

[34] Francisconi-Dos-Rios LF, Calabria MP, Ayco LCC, Honório HM, Carrilho MROC, Pereira JCP, et al. Mint: Chlorhexidine does not improve but preserves bond strength to eroded dentin. American Journal of Dentistry. 2015;**28**:28-32

[35] Breschi L, Mazzoni A, Nato F, Carrilho M, Visintini E, Tjäderhane L, et al. Mint: Chlorhexidine stabilizes the adhesive interface: A 2-year in vitro study. Dental Materials. 2010;**26**:320-325. DOI: 10.1016/j.dental.2009.11.153

[36] Freitas ACC, Ximenes NCA, Aguiar JS, Nascimento SC, Lins TUL, Magalhães LR, et al. Mint: Biological activities of *Libidibia* (*Caesalpinia*) *ferrea* var. parvifolia (Mart. ex Tul.) L. P. Queiroz Pod Preparations. Evidence-based Complementary and Alternative Medicine. 2012;**2012**:514134. DOI: 10.1155/2012/514134

[37] Oliveira AF, Batista JS, Paiva ES, Silva AE, Farias YJMD, Damasceno CAR, et al. Mint: Evaluation of the Brazilian ironwood (*Caesalpinia ferrea* Mart. ex Tul. var. ferrea) healing activity on cutaneous lesions of goats. Revista Brasileira de Plantas Medicinais. 2010;**12**:302-310. DOI: 10.1590/S1516-05722010000300007

[38] Silva LCN, Silva CA, de Souza RM, José Macedo A, da Silva MV, dos Santos Correia MT. Mint: Comparative analysis of the antioxidant and DNA protection capacities of *Anadenanthera colubrina*, *Libidibia ferrea* and *Pityrocarpa moniliformis* fruits. Food and Chemical Toxicology. 2011;**49**:2222-2228. DOI: 10.1016/j.fct.2011.06.019

[39] Guerra ACVA, Soares LAL, Ferreira MRA, Araújo AA, Rocha HAO,

Medeiros JS, et al. *Libidibia ferrea* presents antiproliferative, apoptotic and antioxidant effects in a colorectal cancer cell line. Biomedicine & Pharmacotherapy. 2017;**92**:696-706. DOI: 10.1016/j.biopha.2017.05.123

[40] Zanin JL, de Carvalho BA, Martineli PS, dos Santos MH, Lago JH, Sartorelli P, et al. Mint: The genus *Caesalpinia L.*, (*Caesalpiniaceae*): phytochemical and pharmacological characteristics. Molecules. 2012;**17**:7887-7902. DOI: 10.3390/molecules17077887

[41] Cavalheiro MG, Farias DF, Fernandes GS, Nunes EP, Cavalcanti FS, Vasconcelos IM, et al. Mint: Biological and enzymatic activities of aqueous extract of seeds from *Caesalpinia ferrea* Mart., Leguminosae. Revista Brasileira de Farmacognosia. 2009;**19**:586-591. DOI: 10.1590/S0102-695X2009000400014

[42] Silva FDB, Sales MAG, Sa ORM, Santana GM, De Deus MSM, Castro Souza JM, et al. Mint: Cytotoxic, genotoxic and cytoprotective potential of aqueous extracts of *Caesalpinia pyramidalis* Tul., *Caesalpinia ferrea* Mart. and *Caesalpinia pulcherrima* Sw. Brazilian Journal of Biosciences. 2015;**13**:101-109. Available from: http://www.ufrgs.br/seerbio/ojs/index.php/rbb/article/view/3252

[43] Pereira LP, Silva RO, Bringel PHSF, Silva KES, Assreuy AMS, Pereira MG. Mint: Polysaccharide fractions of *Caesalpinia ferrea* pods: potential anti-inflammatory usage. Journal of Ethnopharmacology. 2012;**139**:642-648. DOI: 10.1016/j.jep.2011.12.012

[44] Conde NCO, Pereira MSV, Bandeira MFCL, Venâncio GN, Oliveira GP, Sampaio FC. Mint: *In vitro* antimicrobial activity of plants of the Amazon on oral biofilm micro-organisms. Revista Odonto Ciência. 2015;**30**:179-183. DOI: 10.15448/1980-6523.2015.4.17794

[45] Oliveira GP, Souza TP, Caetano SK, Farias KS, Venâncio GN, Bandeira MFCL, et al. Mint: Antimicrobial activity in vitro of extracts of the stem bark and fruit of *Libidibia ferrea* l. against microorganisms of the oral cavity. Revista Fitos. 2013;**8**:95-102. DOI: 10.5935/1808-9569.20130004

[46] Marsh PD. Mint: Controlling the oral biofilm with antimicrobials. Journal of Dentistry. 2010;**38**:S11-S15. DOI: 10.1016/S0300- 5712(10)70005-1

[47] Kunle OF, Egharevba HO, Ahmadu PO. Mint: Standardization of herbal medicines—a review. International Journal of Biodiversity and Conservation. 2012;**4**:101-112. DOI: 10.5897/IJBC11.163

[48] Brazil. Ministry of Health. Ordinance no 13 of March 14, 2013. Available from: http://bvsms.saude.gov.br/bvs/saudelegis/anvisa/2013/rdc0013_14_03_2013.html [Accessed: 20-08-2018]

[49] Matos AJP. Physico-chemical quality control of a formulation of orabase ointment of Libidibia ferrea ex. *Caesalpinia ferrea* L [thesis]. Manaus: School of Dentistry, Federal University of Amazonas; 2016

Chapter 7

Innovative Biomaterials for Tissue Engineering

Anna Dolcimascolo, Giovanna Calabrese, Sabrina Conoci and Rosalba Parenti

Abstract

In the field of regenerative medicine, biomaterials play a crucial role since they may serve as a support (scaffold) to promote cell growth and differentiation in order to promote the healing of tissue lesion. The aim of this chapter will be to analyze the properties of more recent biomaterials suitable for tissue engineering strategies, to end to define better and innovative materials for scaffold production. To this purpose, we will analyze the main materials (natural and synthetic) and their characteristics, such as biocompatibility, bioactivity, and biodegradation, and it will be discussed how their chemical-physical properties (surface morphology, porosity, stiffness, and mechanical strength) could affect the interaction with cells and living system. Moreover, the chapter will be focused on methods of extraction or production of biomaterial suitable for scaffolds.

Keywords: biomaterial, scaffold, polymers, biocompatibility, surface, porosity, synthesizing

1. Introduction

The promising field of tissue engineering (TE) purposes to restore damaged tissues by combining cells with biomimetic materials able to act as templates for tissue regeneration and to drive new tissue growth. The term "tissue engineering" was formally conceived at a National Science Foundation workshop in 1988 as "the application of principles and methods of engineering and life sciences toward the fundamental understanding of structure-function relationships in normal and pathological mammalian tissues and the development of biological substitutes to restore, maintain or improve tissue function" [1].

According to the definition of Langer and Vacanti, tissue engineering is "*an interdisciplinary field of research that applies the principles of engineering and the life science toward the development of biological substitutes that restore, maintain or improve tissue function*" [2].

The tissue engineering is a highly multidisciplinary field that associates several areas including clinical medicine, mechanical engineering, materials science, genetics, and related disciplines to both engineering and the life sciences. This field is based principally on the use of biomimetic materials (3D scaffolds) that provide not only a suitable environment for the new developing tissue but also offers a structure for cell adhesion, proliferation, and extracellular matrix (ECM) deposition until new tissue is totally restored [3]. Furthermore, the scaffolds are often combined with cells and signaling molecules or growth factors representing the key elements of tissue engineering.

2. Biomaterials for tissue engineering

The first definition of biomaterial was developed in the 1980s, during the Consensus Development Conference (Chester, UK, 1982) in which the biomaterials were defined as "*any substance, other than a drug, or a combination of substances, synthetic or natural in origin, which can be used for any period of time, as a whole or as a part of a system, which treats, augments or replaces any tissue, organ or function of the body*" [4]. Since ancient times, men searched in nature animal or plant-derived materials able to heal wounds, maintain, or restore body functions. In fact, the ancient Egyptians and Romans used vegetable fibers to sew skin lesions, and they were able to model wooden limb prostheses. Later, the industrial revolution allowed the development of a series of synthetic biomaterials (first metallics and then polymerics) with characteristics more and more suitable for the development of medical devices.

More recently, natural and synthetic biomaterials have become one of the important elements for regenerative medicine and tissue engineering strategies.

Nowadays, several types of scaffolds have been produced with a multiplicity of manufacture systems but the main challenge for tissue engineering is represented from the choice of appropriate materials for the scaffold production. To this aim, different types of biomaterials have been currently used, such as, natural or synthetic polymers, ceramics, metals, composites, and hydrogels. Furthermore, it is important when planning or determining the suitability of a scaffold to evaluate that it fulfills the following key requirements: (i) biocompatibility, (ii) bioactivity, and (iii) biodegradability.

The main requirement of the scaffold for tissue engineering is its biocompatibility or capability to promote cellular adhesion, proliferation, and migration onto the surface and eventually through the scaffold *in vitro* and *in vivo*. Moreover, after implantation, it must integrate into the host tissue without eliciting immune response in order to avoid an important inflammatory reaction that might decline healing or induce rejection [5].

The bioactivity represents the ability of a biomaterial to interact with surrounding tissue ensuring cell adhesion, proliferation, and differentiation [6]. Generally, biomaterials with chemical composition comparable to the host tissue have a higher bioactivity and can promote cellular recognition evoking specific cellular response to support tissue growth. To this aim, it is possible to modify the surface of the biomaterial by adding extracellular matrix macromolecules, including collagen, fibronectin, and laminin, to produce a biomimetic environment equivalent to the native tissue able to modulate cellular behavior and response [7].

On other essential property of scaffold for tissue engineering is the biodegradability. The biomimetic scaffolds are not permanent implants but they must be biodegradable to allow cells to produce their own extracellular matrix. Further, the by-products of this degradation must also be nontoxic and easily eliminable from the body without interfering with other tissues [8]. On the other hand, it is critically important to also know the in vivo degradation kinetics of a biomaterial to avoid an excessively rapid or slow elimination. In the first case, the scaffold could not satisfy its function of support for cells, while in the second one, it could cause necrosis or inflammation [9].

3. Scaffold design: the importance of structural and mechanical properties

The scaffold for tissue engineering must possess structural and mechanical characteristics appropriate to the anatomical site in which it must be implanted and, moreover, must be strong enough to allow its surgical manipulation during implantation. The structural features include macro- and micro-structural properties. The macro-structural properties refer to a temporary 3D architecture, of critical importance, which mimic the ECM and allows cell to maintain their native differentiated phenotypes; while, the micro-structural properties refer to scaffold porosity, pore shape, pore size, and interconnectivity. The mechanical properties include mechanical strength and stiffness.

3.1 Structural properties

Scaffold micro and macro architecture critically influences cell survive and surface adhesion, but also cellular proliferation, differentiation, vascularization, and specific gene expression [10].

If on the one hand, a scaffold may be strong enough to support the physiological load of the body and to allow surgical handling during implantation, on the other hand, it is important to obtain a porous structure to avoid cellular colonization. It is clear that a balance between mechanical strength and high porosity is a significant challenge in scaffold production.

3.1.1 Pore interconnection, porosity, and pore size

Pore interconnection, porosity, and pore size represent very important parameters for the scaffold production. All three features allow cellular penetration, vascularization, adequate diffusion of nutrients and oxygen to cells within the construct, and neo-formed extracellular matrix ensuring cell viability [5, 11].

In particular, pore size is a key element for the scaffold efficiency. In fact, the pores must be large enough to allow cells to penetrate and migrate within the scaffold structure, but also small enough to allow the binding of a critical number of cells at the same. Pores can be classified into micropores (0.1–2 nm), mesopores (2–50 nm), and macropores (>50 nm) according to their dimension. All the scaffolds used for tissue engineering may have a macroporous structure with a specific pore size as a function of the type of host tissue. In particular, a pore size of 20 micron is required for hepatocyte and fibroblast growth, while the dimension is around 20–150 micron for soft tissue healing. For bone tissue engineering, researchers propose a pore size range between 200 and 400 micron.

The most common techniques used to obtain a porous structure are gas foaming, salt leaching, phase separation, sintering, and freeze-drying.

3.2 Mechanical properties

The scaffold for tissue engineering must have adequate mechanical integrity, so that it can offer support from the time of implantation until the remodeling process is fully completed.

3.2.1 Mechanical strength and stiffness

The mechanical strength depends on the bonding forces that hold together the atoms in scaffold architecture. It is an important parameter to avoid the solid structure deformation due to cellular loading on the scaffold or caused by scaffold handling.

Another important feature of the scaffold surface is the stiffness that is measured by Young's modulus. Cells respond to scaffold stiffness *via* different mechanisms such as activation of ion channels or protein unfolding, and by this way, stiffness affects cell proliferation and differentiation. Hadjipanayi *et al.* demonstrated that the increasing of free-floating collagen matrix stiffness led to a higher proliferation rate for human dermal fibroblasts [12].

4. Biomaterials for scaffold production

Biomaterials for tissue engineering have a considerable importance for the success of a tissue replacement or regeneration. In addition to interacting with the implant site, they have the ability to influence biological processes that are important for tissue regeneration.

Different kinds of biomaterials have been used for scaffold production such as ceramics and polymers, naturals and synthetics, metals, composites, and hydrogels.

4.1 Ceramics

For several decades, ceramic biomaterials have been used to reconstruct damaged body parts and for skeletal repair.

Ceramic biomaterials are inorganic compounds of natural or synthetic origin, which may contain metallic and nonmetallic elements. These biomaterials are generally made of polycrystalline solids, rarely of monocrystals and sometimes have an amorphous structure. Generally, their mechanical properties, including hard surface, high mechanical stiffness, low elasticity, low thermal expansion, chemical-physic refractoriness, depend on the way they are produced or extracted, but their properties can also depend on the composition and particle size of the starting powders.

Ceramic scaffolds are commonly used for bone regeneration practices because they are highly biocompatible, rarely evoke an immune response, and hardly cause the formation of fibrous tissue around the scaffold; instead they are osteoinductive, considering their high ability to recruit cells from the biological environment and promote osteogenic differentiation. Although the ceramics present these advantages, their use in tissue engineering applications is limited due to their fragility and slow degradation [1, 13, 14].

On the basis of their main features, they can be distinguished into three categories: (a) bio-inert ceramics: completely inert to biological environment; (b) resorbable materials: subjects to *in vivo* degradation for phagocytosis or dissolution of the material in biological fluids; and (c) bioactive ceramics: able to form chemical bound with the cell surface [15].

The most common ceramic biomaterials used for tissue regeneration are: (1) CaP, including hydroxyapatite (HA) ($Ca_{10}[PO_4]_6[OH]_2$), beta-tricalcium phosphate (BTF) ($Ca_3[PO_4]_2$), biphasic calcium phosphate (mixture of hydroxyapatite and beta-tricalcium phosphate), (2) bioglass, (3) alumina (Al_2O_3), and (4) zirconia oxide (ZrO_2).

CaP biomaterials are often selected for bone graft since they mimic bone tissue composition. One of the first used ceramic biomaterials for skeletal repair was BTF in 1920 by Albee and Morrinson [16].

HA may be natural or synthetic. Natural HA derives from particular species of coral or bovine bone and can contain traces of other elements such as Mg, Na,

CO_3, and F. Synthetic HA is prepared by sintering in dense or macroporous form as granules or blocks [17]. Ray and Ward, first, showed the high biocompatibility and biomimicry of synthetic HA in their study in which they used this material for bone tissue engineering application in the long bones and iliac wings of dogs [18]. Later, numerous other studies on HA have been carried out. Calabrese *et al.* in their studies tested a composite bi-layer type-1 collagen-HA/Mg scaffold for osteochondral regeneration, both *in vitro* and *in vivo*. They showed that the combination of this scaffold with mesenchymal stem cells (MSC) derived from adipose tissue (hAD-SCs) in the presence of specific differentiation conditions induce osteochondro differentiation both *in vitro* and *in vivo* [19–23].

Bioglass is composed by 45 wt% SiO_2, 24.5 wt% CaO, 24.5 wt% Na_2O, and 6.0 wt% P_2O_5. The first one (45S5 Bioglass) has been developed by Hench, which used it for biomedical applications *in vitro* and *in vivo* [24].

Bioglass materials can be synthesized through different methods such as polymer foam replication, thermal bonding of particles or fibers, and sol-gel processing. Similarly to HA, it is suitable for bone graft due to the high ratio of calcium to phosphorus promoting the formation of apatite crystals on its surface after grafting. Bioglass materials offer high osteoinductivity, control of rate of degradation, and excellent bioactivity even if they can present poor mechanical properties such as low strength and toughness [25, 26].

Alumina (Al_2O_3) is a ceramic biomaterial with a crystalline structure. Generally, a low porosity and reduced grain size increase its mechanical strength. Like other ceramic materials, alumina is fragile but it has good tribological properties such as resistance to wear.

Zirconia is characterized by a polymorphic structure and offer has a hard surface, a low thermal conductivity, and a high coefficient of thermal expansion. Its excellent biocompatibility and high breaking load make it a good candidate for prosthesis and bone grafting.

4.2 Polymers

Various biological polymers such as collagen, alginate, proteoglycans, chitin, and chitosan have been used to produce scaffolds. They are biocompatible and bioactive promoting cellular adhesion and growth on their surface. However, they often show poor mechanical properties and fast biodegradability, which limit their use.

Collagen and its derivatives are good candidates for osteochondral regeneration but also tendon and ligament reconstruction since the extracellular matrix of these tissues is mostly made of type-1 collagen fibers [27–29].

Collagen scaffolds are highly bioactive ensuring excellent cellular adhesion to their surface. However, since they have low resistance to mechanical stress often are coupled with other materials, which improve their mechanical properties.

Several studies have been focused on the use of collagen scaffolds for tissue engineering strategies. Aravamudhan *et al.*, for example, reported the fabrication and characterization of cellulose and collagen-based micro-nanostructured scaffolds exhibiting mechanical features similar to those of trabecular bone that promoted good adhesion of human osteoblasts to their surface. Moreover, they underwent a progressive calcium deposition process compared to control polyester micro-nanostructured scaffolds [30].

In another study, Schneider *et al.* developed a collagen I/III hydrogel scaffold and used it to seed hMSC isolated from bone marrow of femoral head spongiosa and from umbilical cord. When stimulated with osteogenic induction medium, both cell types showed comparable osteogenic gene expression, migration, and scaffold colonization [31]. Collagen scaffolds may also be used to deliver osteogenic differentiation factors as demonstrated by Lu H. *et al,* who immobilized BMP4 in a collagen-PLGA hybrid scaffold to promote osteogenesis [32].

Polysaccharides such as chitin, chitosan, and alginate are suitable for both hard and soft tissue regeneration. In particular, chitosan scaffolds can be manufactured by freeze-drying techniques, which allow obtaining a porous scaffold with high pore interconnectivity. Chitosan ensures good cellular adhesion and thank to its positive charges can interact with glycosaminoglycans and proteoglycans present in living tissues. Costa-Pinto *et al.* cultured human bone marrow MSC on melt-based porous chitosan scaffolds using an osteogenic differentiation medium. They found an increase of cell viability and ALP activity after 21 days. They also investigated the capacity of the cell seeded scaffold to repair a cranial defect in mouse, and 8 weeks after implantation bone formation in the scaffold was analyzed using Bone μCT [33]. Chitosan may also be used as an injectable biomaterial as demonstrated by Bi *et al.,* who produced a composite scaffold of tricalcium phosphate (TCP), chitosan, and platelet rich plasma (PRP). MSC seeded on injectable biomaterial was used in vivo to test its capacity to repair bone fracture in goat femora [34].

Synthetic polymers are high molecular weight compounds composed of a series of monomeric units. On the basis of their structure, they can be linear, branched, or cross-linked. Considering their thermo-mechanical properties, they are thermoplastic or thermosetting. Polymeric materials can be produced in the form of fibers, films, bars, and viscous liquids, and they offer the important advantage to modulate their mechanical properties and biodegradation by varying synthesis process and reactants used. However, they could have low biocompatibility and mechanical strength and show *in vivo* toxicity due to the release of ions and other residual particles of polymerization.

Among the different synthetic polymers, the most suitable for scaffold production is the bio-erodible. These kinds of polymers undergo surface degradation with production of nontoxic low molecular weight compounds.

Numerous synthetic polymers have already been used such as: polystyrene, thermoplastic aromatic polymer with a linear structure; poly-l-lactic acid (PLA), hydrophobic polymer with slow degradation rate due to microorganisms; polyglycolic acid (PGA), hydrophilic polymer with good mechanical properties and fast degradation; poly-dl-lactic-co-glycolic acid (PLGA), biocompatible copolymers with fast degradation rate; and polycaprolactone (PCL), highly hydrophobic polymer with good permeability.

In particular, PGA and PLA and their copolymers are natural polyesters normally present in the organism and therefore well tolerated. They have been used for suture threads, orthopedic screws, and prostheses manufacture since 1970, and more recently, they have been evaluated for scaffold production and tissue engineering strategies. About this, Eğri *et al.* combined PLA and PGA to obtain a PLA-PEG-PLA scaffolds able to release VEGF and BMP-2 in bone tissue lesion. In relation to its chemical composition, the scaffold allows fast release of VEGF in about 1 week and slower constant release of BMP-2 [35].

4.3 Metals

Metals are particularly suitable for tissue engineering strategies for their good mechanical properties such as high elastic module, yield strength, and high ductility allowing them to bear a load without being deformed. If mechanical resistance makes them excellent candidates for scaffold production, however, the reduced cell adhesion to their surface could be a considerable limit to their use. Moreover, metal implants can release toxic metallic ions and/or particles, and biological fluids can show corrosive action on their surface that can alter their function. Among the different metals used for scaffold production, there are stainless steel, cobalt, and titanium alloys.

Stainless steels are iron-based alloys with a low content of carbon and a high content of chromium. The presence of carbon ensures good mechanical properties but determines carbides formation that makes the scaffold subject to corrosion in a biological environment.

Cobalt-based alloys are of two types: cobalt/chromium/molybdenum alloy obtained with casting/melting methods and cobalt/nickel/chromium/molybdenum alloy worked by forging. Generally, the high level of chromium and molybdenum typical of these alloys increase granule size and improve mechanical properties.

Titanium alloys can be *alpha, beta or alpha/beta* biphasic. *Alpha* alloys contain alpha stabilizers such as aluminum and gallium and are characterized by good strength, hardness, resistance sliding, and weld ability; *Beta* alloys contain beta stabilizers such as vanadium, niobium, and tantalus molybdenum and show good ductility. *Alpha/beta* biphasic alloys show a mix of *alpha/beta* stabilizers, and they are quite ductile even if little resistant to high temperatures, and the most suitable one for biomedical application is Ti 6Al 4 V.

Wohlfahrt *et al.* tested the osteoinductivity and osteointegration capabilities of Ti and TiO2 scaffolds in rabbit tibia peri-implant osseous defects. After 4 weeks, the implant was removed and the new bone formation was observed. Moreover, a gene expression analysis was performed considering different osteogenesis differentiation markers such as osteocalcin and collagen-I [36].

In another study, Zuchuat *et al.* developed Cr-Co-Mo membranes and placed them in rabbit tibiae to analyze the volume of new bone formation. After the explant, histological analysis showed a huge number of osteoblasts and osteocytes on the scaffold [37].

4.4 Composites

Composite scaffolds are developed combining different biomaterials such as natural or synthetic polymers (PGA, PLA, gelatin, chitin, and chitosan), ceramics (hydroxyapatite and beta-tricalcium phosphate or bioglasses), and metals. They have technological, industrial, and applicative importance since they combine biocompatibility, biodegradation, and appreciable mechanical strength. Moreover, these kinds of scaffolds could be applied for both hard and soft tissue regeneration and greatly mimic tissue architecture being composed of cells and extracellular matrix.

Several studies displayed the efficacy of composite scaffolds (polymers/ceramics and synthetic/natural polymers) for tissue engineering strategies [38, 39].

Other researchers demonstrated that another interesting solution may be the combination of metallic implants with polymer coating or metal/ceramic scaffolds [40, 41].

4.5 Hydrogels

Hydrogels are hydrophilic polymers rich of polar moieties such as carboxyl, amide, amino, and hydroxyl groups, held together by chemical bounds or physical intra-molecular and inter-molecular attractions. Their main feature is the ability to absorb enormous amounts of water or biological fluids and swell without dissolving.

According to their origin, hydrogel can be classified into natural (made of polypeptides and polysaccharides), synthetic (obtained by traditional polymerization),

	Advantages	Disadvantages	Clinical uses
Ceramics	-Hard surface -High mechanical stiffnes -Chemical-physic refractoriness -High biocompatibility -osteoinductivity	- Brittleness - Slow degradation - Processing difficulties	-Hip prosthesis -Dental prosthesis -Bone and cartilage
Natural polymers	-Biocompatibility -Bioactivity	- Poor mechanical properties - Fast biodegradation	-Bone and cartilage - Tendon and ligament n
Synthetic polymers	- Possibility of modulating porosity and mechanical properties during the synthesis process.	- Low biocompatibility: possible release of ions and other residual particles of polymerization - Low mechanical strength	-Sutures -Catheters -Cardiovascular prostheses -Bone cements
Metals	-Good mechanical properties: high elastic module, yield strength and high ductility	- Reduced cell adhesion to their surface - Possible corrosion mediated by biological fluid	-Dentistry and orthopedic prostheses
Composites	-Biocompatibility -Good mechanical properties	- Processing difficulties	-Hard and soft tissue
Hydrogel	- Biocompatibility - Controlled biodegradation in vivo -Possibility to modulate their parameters [cross-linking density, porosity, pore size and interconnectivity]		- Hard and soft tissue

Table 1.
Advantages, disadvantages, and main clinical uses of different kinds of biomaterials.

and semi-synthetic. Moreover, they can present an amorphous or semi-crystalline structure that can be cationic, anionic, neutral, or ampholytic. Depending on their stability in a biological system, they can be considered durable if they do not undergo chemical-physical modification or biodegradable if they degrade into oligomers, which are subsequently eliminated from the body. In the last decades, smart hydrogels have been developed featured by the possibility to modify their structure and mechanical properties according to environmental stimuli such as pH or temperature. Already 50 years ago, these materials have been appreciated for their chemical-physical characteristics by Wichterle and Lim, who developed a poly(2-hydroxyethyl methacrylate)-based hydrogel for contact lens [42]. Since they present a soft and rubbery consistency very similar to that of ECM of different tissues, they have been recently studied for tissue engineering strategies. In particular, hydrogels used for scaffold production may respond to important requirements such as biocompatibility and controlled *in vivo* biodegradation. It is very important to modulate parameters such as hydrogel cross-linking density, porosity, pore size, and interconnectivity to obtain a suitable structural for cellular colonization and proliferation. Hydrogels can be modified at the surface by peptides or growth factor, which can promote cell attachment and differentiation process. Generally, natural hydrogels are less toxic and more tolerated than synthetic ones, and Pasqui *et al*, for example, developed a natural cellulose-hydroxyapatite hybrid hydrogel for bone tissue engineering. For the chemical synthesis procedure, the freeze-dried hydrogel was immersed in a solution containing HA microcrystals, and then an *in vitro* study demonstrated that MG63 osteoblast-like human cell seeded into hydrogel samples adhered and proliferated rapidly. Moreover, an increase of ALP activity was identified at 3, 7, and 14 days [43]. Synthetic hydrogels could have limitations in the biocompatibility, but they offer the possibility to modulate their mechanical features and rate of degradation in biological environment. Kinard *et al.* developed a biodegradable oligo[poly(ethylene glycol)fumarate] hydrogel to deliver demineralized bone matrix (DBM) in a rat bone defect. They found that the *in vivo* degradation rate of the hydrogel depend on the DBM content, higher was the rate of DBM faster was the degradation. Moreover, high content of DBM could affect the mechanical properties of the hydrogel even if it increases its osteoinductivity in vitro and in vivo [44] (**Table 1**).

5. Processing techniques for scaffold production

After the choice of the biomaterial to use for scaffold production, it is quite important to select an adequate processing technique that allows to maintain high levels of control of the macro- and micro-structural properties of the same. The processing methodology must satisfy key requirements such as: process accuracy and repeatability. The scaffolds obtained will present regular shaped pores with consistent pore size and interconnectivity and should not show any physical-chemical variations when produced by the same method. Moreover, the processing conditions must not alter the mechanical properties of the biomaterial, and any toxic solvent used during the process must be totally removed not to limit scaffold clinical use [3, 11]. Among the most spread processing techniques, probably the most known are those that foresee the employment of a porogenous organic or inorganic agent such as sodium chloride, sodium tartrate, sodium citrate, citric acid, or saccharose. However, the use of porogens limits the scaffolds to thin membranes with a thickness of 2 mm to facilitate complete porogen removal [45].

Mikos *et al.* described solvent casting/particulate leaching for the first time, and it is chosen for the fabrication of porous scaffold used for bone tissue engineering. In this case, the porous agent is dispersed in appropriate solvent and then the dispersion is processed by casting or by freeze-drying. This technique allows obtaining thin membranes with 30–300 micrometer pore size and 20–50% porosity even if the pores have a shifting shape and the interconnectivity is quite low. However, the method presents some disadvantages like time consuming (it is necessary to wait for days or weeks for solvent evaporation) and the use of toxic organic solvents [46].

In melt molding/particulate leaching, an unrefined thermoplastic polymer is mixed with the porous agent and then the blend is poured in a mold with an appropriate shape. The mold is then heated above the glass transition temperature of the polymer and at last the obtained solid is immersed in a solvent to promote the dissolution of the porogens. The advantage of this methodology is the possibility to monitor the pore size and porosity (generally 80–84%) by varying the amount of porogenous [47]. A good variant of melt molding is extrusion or injection molding proposed by Gomes *et al.*, who replaced the porous agent with a blowing agent based on citric acid. During the heating process, the blowing agent degraded producing carbon dioxide which formed interconnected and well-shaped pores [48].

Gas foaming is an high pressure processing technique described by Mooney *et al* who produced sponges of poly(D,L-lactic-co-glycolic acid) without the use of organic solvents. Solid disks of the polymer are exposed to high pressure CO_2

(5.5 MPa) at room temperature followed by a decreasing of gas pressure to reduce its solubility in the polymer bulk. It brings CO_2 to abandon the polymer forming well-shaped pores [49].

Phase inversion/particulate leaching is a valid method to obtain polymeric scaffolds. After the polymer solubilization in a suitable solvent, the solution is dissolved in water that provokes the polymer precipitation. Obviously, it is possible to modulate the characteristics of the scaffolds obtained through this method by varying the polymer concentration but also the temperature of the solution. Holy *et al.* used this technique to develop porous PLGA scaffold with architecture similar to osseous trabecular for bone tissue engineering [50].

Another interesting method is the fiber bonding. It allows obtaining scaffolds containing a dense frame of synthetic fibers that form a sufficiently porous three-dimensional structure. This technique provides the alignment of the PGA fibers in the desired orientation and subsequently they are covered with a PLLA/methylene chloride solution and heated above the melting temperatures of both polymers. When PLLA is removed through a dissolution process, the PGA fibers remain attached to each other forming a thick net.

In the freeze-drying method, the polymer solution is first frozen rapidly at temperatures below 0°C followed by solvent removal by vacuum sublimation. It can be applied to obtain both natural and synthetic scaffolds [51]. At last, the progress of computer technology led to the development of new techniques like solid freeform fabrication (SSF) whose introduction has signed a new era for manufacturing industry. These techniques allow to produce layer-by-layer 3D objects starting from information generated by CAD system or computer-based medical imaging modalities. Obviously, the use of a computerized production system saves time and modulates with extreme precision parameters related to the micro and macro architecture of the scaffold.

The first SFF technique used for tissue engineering purpose was 3D printing. This technique uses a printer head that places a liquid binder onto thin layers of powder following the object shape generated by a CAD system. Using this technique, Kim *et al.* obtained porous PLGA scaffolds [52], while Zeltinger *et al.* created poly(l-lactic acid) disk shaped scaffolds with two different porosities (75% and 90%) and four different pore size distributions (<38, 38–63, 63–106, and 106–150 μm) [53].

Another interesting SFF methodology is fused deposition modeling (FDM). In this case, a filament of thermoplastic material is fed and melted inside a heated liquifier head and then it is forced out by an extruder and deposited on a platform. Layer by layer, the 3D object is then obtained. By varying the direction of material deposition for each layer, it is possible to change the pore size and interconnectivity of the scaffold. Using this methodology, Hutmacher et al. obtained polycaprolactone scaffolds with honeycomb-like structure and a porosity of 61 +/− 1% and proved their in vitro ability to promote proliferation of primary human fibroblasts and periosteal cells [54].

6. Biomaterials for bone tissue engineering: current applications and new perspectives

One of the current problems in orthopedic clinic is represented by bone lesions caused by traumas, cancer resection degenerative diseases, or nonunion of fractures, which do not heal spontaneously but require surgical procedures. Today, the gold standard for osseous replacement is the autologous bone graft. This technique employs cells of the same patient generally taken from different sites such as fibula or iliac crest that are implanted in bone defect to promote a rapid healing. Although it minimizes the risk of autoimmune response, which was the critical side of xenogenic grafts, it presents some disadvantages such as donor site morbidity, infections, and post-surgery chronic pain [55–57]. In sight of this, science aims to find innovative solutions like application of biomaterials to orthopedics in order to develop medical implant useful to accelerate the healing, restoring the physiological functions of bone.

The design of an implant for skeletal defects may consider the main characteristics of bone tissue which is divided into two different forms: cortical bone, almost solid with less than 10% porosity and trabecular bone organized in a sponge-like pattern with a porosity of 50–60% [58]. According to the classification of Hanch and Navarro, the evolution of bone implant devices has marked three different generations: (a) bio-inert materials (first generation), (b) bioactive and biodegradable materials (second generation), and (c) biomaterials capable of inducing specific cellular responses by incorporating into a 3D scaffold bone progenitor cells and growth factors [59, 60].

The purpose of first generation of implants was the integration with host tissue without eliciting specific immune response. These implants include metals (Stainless steel, Ti-based, and Co–Cr-based alloys), ceramics (natural and synthetic HA), and polymers (silicone rubber, PE, acrylic resins, polyurethanes, polypropylene PP, and polymethylmethacrylate).

The second generation of implants was developed between 1980 and 2000 and intends to improve both the bioactivity and *in vivo* biodegradation. To this purpose, one of the possible strategies was to modify the first generation by strategic coating such as HA, *beta*-tricalcium phosphate (*b*-TCP), or bioactive glass. Another innovation was the use of natural or synthetic polymers like, poly(*p*-caprolactone), polylactide, polyglicolide, and chitosan with controlled *in vivo* biodegradation rate.

Third generation of implants combines biomaterials useful to develop 3D porous bioactive, biodegradable scaffolds with the integration of progenitor cells, and specific growth factors. This innovation laid the foundations for modern bone tissue engineering strategies. Even if an ideal combination of biomaterials for scaffold production has not been identified yet, recent studies have demonstrated the great efficiency of ceramics in mimic chemical-physic characteristics of bone tissue ECM. Also, our group tested *in vitro* and *in vivo* potential of collagen type-1/Ha-Mg combination to promote bone injury healing. We demonstrated that although biomimetic scaffolds are "*per sè*" able to promote tissue regeneration thanks to their high osteoinductivity, their combination with progenitor cells and growth factors would be more efficient [19–21]. Generally, osteogenic cells such as adult stem cells (ASC) isolated from adult tissues like bone marrow, adipose tissue, or muscle are good candidate to be transplanted in skeletal lesion together with an appropriate scaffold. These kinds of cells are characterized by high capacity of self-renewal and potential of osteogenic differentiation. Moreover, it has been suggested that ASC possess immunosuppressive effects, which make them particularly privileged for transplantation in vivo.

Growth factors are cytokines normally secreted by different cell types. Acting on their own receptors, they induce intracellular pathways, which promote proliferation, cellular adhesion, and differentiation. Bone tissue produces different growth factors such as bone morphogenetic proteins (BMPs), transforming growth factor beta (TGFb), fibroblast growth factors (FGFs), insulin growth factor I and II (IGF I/II), and platelet-derived growth factor (PDGF), which have been proposed for tissue engineering strategies. In particular, BMP2 and 7 have been cloned and are commercially available as recombinant proteins. The interest in them for bone regenerative practices has increased since 1965 when Urist discovered that demineralized bone transplanted in subcutaneous tissue induces bone formation [61]. This potential was later attributed to the presence of

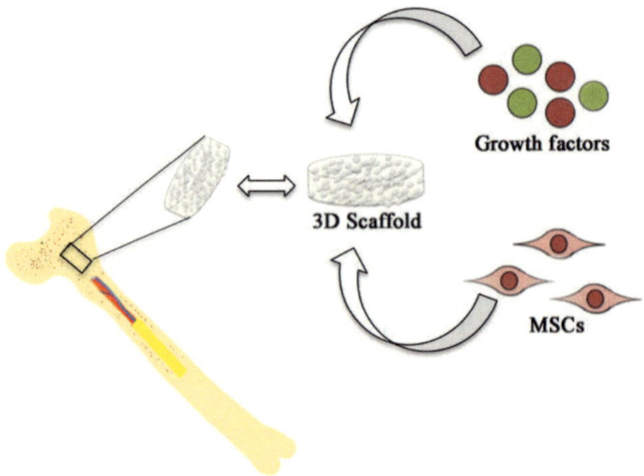

Figure 1.
Schematic representation of bone tissue engineering.

BMP. Obviously, the choice to include a growth factor in the scaffold requires the use of biomaterials that can act as drug delivery systems protecting the cytokine from *in vivo* proteolysis and ensuring a progressive and controlled release over time. In this sense, a good alternative is the physical immobilization of the growth factor in a biodegradable hydrogel. In this case, the release will be controlled by the *in vivo* degradation of hydrogel cross-linked with generation of water-soluble hydrogel fragments [9] (**Figure 1**).

7. Conclusions

Recently, the interest in natural and synthetic biomaterials for medical devices production has increased, and more and more in-depth studies are carried out to better detect their possible applications linked to chemical-physical features and the extractive or synthetic methods, which do not alter their structural properties and biocompatibility. Moreover since tissue engineering strategies have become a valid alternative for body structure and function restoring, biomaterials are also used for the fabrication of 3D porous biomimetic bioactive scaffolds with controlled degradation rate *in vivo.*

As previously mentioned, the main classes of biomaterials for scaffold production are ceramics, natural and synthetic polymers, metals, composites, and hydrogels. *In vitro* and *in vivo* studies have showed the advantages related to their use in regenerative medicine field but they have also highlighted the possible negative sides.

Regarding the application of biomaterials to tissue engineering, the current aim of science is to find the natural or synthetic substance or the combination with the most satisfactory performance *in vivo,* able to promote cell proliferation and differentiation in a tissue lesion in order to restore the normal architecture of ECM.

In conclusion, tissue engineering strategies especially in orthopedic clinic field represent an effective and sophisticated alternative for the future, but their success strictly depends on an ever deeper knowledge about the characteristics of the biomaterials and the potentialities of their combinations.

Conflicts of interest

The authors declare no conflict of interest.

Author details

Anna Dolcimascolo[1], Giovanna Calabrese[1*], Sabrina Conoci[2] and Rosalba Parenti[1]

1 Department of Biomedical and Biotechnological Sciences, Physiology Section, University of Catania, Catania, Italy

2 STMicroelectronics Stradale Primosole, Catania, Italy

*Address all correspondence to: soniacalabrese@hotmail.com

References

[1] O'Brien FJ. Biomaterials & scaffolds for tissue engineering. Materials Today. 2011;**14**(3):88-95. DOI: 10.1016/S1369-7021(11)70058-X

[2] Langer R, Vacanti JP. Tissue engineering. Science. 1993;**260**(5110): 920-926

[3] Salgado AJ, Coutinho OP, Rui L, Reis RL. Bone tissue engineering: State of the art and future trends. Macromolecular Bioscience. 2004;**4**:743-765. DOI: 10.1002/mabi.200400026

[4] Ravaglioli A, Krawjesky R. Bioceramics. Netherlands: Springer; 1992. ISBN 978-94-010-5032-6

[5] Hutmacher DW. Scaffolds in tissue engineering bone and cartilage. Biomaterials. 2000:175-189. DOI: 10.1016/B978-008045154-1.50021-6

[6] Yu X, Tang X, Gohil SV, Laurencin CT. Biomaterials for bone regenerative engineering. Advanced Healthcare Materials. 2015;**4**(9):1268-1285. DOI: 10.1002/adhm.201400760

[7] Tirrell M, Kokkoli E, Biesalski M. The role of surface science in bioengineered materials. Surface Science. 2002;**500**:61-83. DOI: 10.1016/S0039-6028(01)01548-5

[8] Babensee JE, Anderson JM, McIntire LV, Mikos AG. Host response to tissue engineered devices. Advanced Drug Delivery Reviews. 1998;**33**:111-139

[9] Tabata Y. Biomaterial technology for tissue engineering applications. Journal of the Royal Society Interface. 2009;**6**:S311, S324. DOI: 10.1098/rsif.2008.0448

[10] Ratner B, Hoffman A, Schoen F, Jack Lemons J. Biomaterials Science, 3rd Edition. An Introduction to Materials in Medicine. Academic Press; 2012.

[11] Leong KF, Cheah CM, Chua CK. Solid freeform fabrication of three-dimensional scaffolds for engineering replacement tissues and organs. Biomaterials. 2003;**24**(13):2363-2378

[12] Hadjipanayi E, Mudera V, Brown RA. Close dependence of fibroblast proliferation on collagen scaffold matrix stiffness. Tissue Engineering and Regenerative Medicine. 2009;**3**(2): 77-84. DOI: 10.1002/term.136

[13] Lichte P, Pape HC, Pufe T, Kobbe P, Fischer H. Scaffolds for bone healing: Concepts, materials and evidence. Injury. 2011;**42**(6):569-573. DOI: 10.1016/j.injury.2011.03.033

[14] Bose S, Roy M, Bandyopadhyay A. Recent advances in bone tissue engineering scaffolds. Trends in Biotechnology. 2012;(10):546-554. DOI: 10.1016/j.tibtech.2012.07.005

[15] Best SM, Porter AE, Thian ES, Huang J. Bioceramics: Past, present and for the future. Journal of the European Ceramic Society. 2008;**28**:1319-1327. DOI: DOI :10.1016/j.jeurceramsoc.2007.12.001

[16] Albee F, Morrison H. Studies in bone growth. Annals of Surgery. 1920;**71**:32-38

[17] Lagers RZ. Properties of osteoconductive biomaterials: Calcium phosphates. Clinical Orthopaedics and Related Research. 2002;**395**:81-98

[18] Ray RD, Ward AA. A preliminary report on studies of basic calcium phosphate in bone replacement. In: Surgical Forum, American College of Surgeons, 1951. Philadelphia: WB Saunders Co; 1952. pp. 429-434

[19] Calabrese G, Giuffrida R, Fabbi C, Figallo E, Furno DL, Gulino R, et al. Collagen-hydroxyapatite scaffolds

induce human adipose derived stem cells osteogenic differentiation in vitro. PLoS One. 2016;**11**(3). DOI: 10.1371/journal.pone.0151181

[20] Calabrese G, Giuffrida R, Forte S, Salvatorelli L, Fabbi C, Figallo E, et al. Bone augmentation after ectopic implantation of a cell-free collagen-hydroxyapatite scaffold in the mouse. Scientific Reports. 2016;**6**:36399. DOI: 10.1038/srep36399

[21] Calabrese G, Giuffrida R, Forte S, Fabbin C, Figallo E, Salvatorelli L, et al. Human adipose-derived mesenchymal stem cells seeded into a collagen-hydroxyapatite scaffold promote bone augmentation after implantation in the mouse. Scientific Reports. 2017;**7**(1):7110. DOI: 10.1038/s41598-017-07672-0

[22] Calabrese G, Forte S, Gulino R, Cefalì F, Figallo E, Salvatorelli L, et al. Combination of collagen-based scaffold and bioactive factors induces adipose-derived mesenchymal stem cells chondrogenic differentiation in vitro. Frontiers in Physiology. 2017;**8**(50). DOI: 10.3389/fphys.2017.00050

[23] Calabrese G, Gulino R, Giuffrida R, Forte S, Figallo E, Claudia Fabbi C, et al. In vivo evaluation of biocompatibility and chondrogenic potential of a cell-free collagen-based scaffold. Frontiers in Physiology. 2017;**8**:984. DOI: 10.3389/fphys.2017.00984

[24] Hench LL. The story of bioglass. Journal of Materials Science. Materials in Medicine. 2006;**17**(11):967-978. DOI: 10.1007/s10856-006-0432-z

[25] Rahaman MN, Day DE, Bal BS, Fu Q, Jung SB, Bonewald LF, et al. Bioactive glass in tissue engineering. Acta Biomaterialia. 2011;**7**(6):2355-2373. DOI: 10.1016/j.actbio.2011.03.016

[26] Rezwan K, Chen QZ, Blaker JJ, Boccaccini AR. Biodegradable and bioactive porous polymer/inorganic composite scaffolds for bone tissue engineering. Biomaterials. 2006;**27**(18):3413-3431. DOI: 10.1016/j.biomaterials.2006.01.039

[27] Vagaská B, Bacáková L, Filová E, Balík K. Osteogenic cells on bio-inspired materials for bone tissue engineering. Physiological Research. 2009;**59**(3):309-322

[28] Deponti D, Di Giancamillo A, Gervaso F, Domenicucci M, Domeneghini C, Sannino A, et al. Collagen scaffold for cartilage tissue engineering: The benefit of fibrin glue and the proper culture time in an infant cartilage model. Tissue Engineering Part A;**20**(5-6):1113-1126. DOI: 10.1089/ten.TEA.2013.0171

[29] Hoyer M, Drechsel N, Meyer M, Meier C, Hinüber C, Breier A, et al. Embroidered polymer-collagen hybrid scaffold variants for ligament tissue engineering. Materials Science & Engineering. C, Materials for Biological Applications. 2014;**43**:290-299. DOI: 10.1016/j.msec.2014.07.010

[30] Aravamudhan A, Ramos DM, Nip J, Harmon MD, James R, Deng M, et al. Cellulose and collagen derived micro-nano structured scaffolds for bone tissue engineering. Journal of Biomedical Nanotechnology. 2013;**9**(4):719-731

[31] Schneider RK, Puellen A, Kramann R, Raupach K, Bornemann J, Knuechel R, et al. The osteogenic differentiation of adult bone marrow and perinatal umbilical mesenchymal stem cells and matrix remodelling in three-dimensional collagen scaffolds. Biomaterials. 2010;**31**(3):46780. DOI: 10.1016/j.biomaterials.2009.09.059

[32] Lu H, Kawazoe N, Kitajima T, Myoken Y, Tomita M, Umezawa A,

et al. Spatial immobilization of bone morphogenetic protein-4 in a collagen-PLGA hybrid scaffold for enhanced osteoinductivity. Biomaterials. 2012;**33**(26):6140-6146. DOI: 10.1016/j.biomaterials.2012.05.038

[33] Costa-Pinto AR, Correlo VM, Sol PC, Bhattacharya M, Charbord P, Delorme B, et al. Osteogenic differentiation of human bone marrow mesenchymal stem cells seeded on melt based chitosan scaffolds for bone tissue engineering applications. Biomacromolecules. 2009;**10**(8):2067-2073. DOI: 10.1021/bm9000102

[34] Bi L, Cheng W, Fan H, Pei G. Reconstruction of goat tibial defects using an injectable tricalcium phosphate/chitosan in combination with autologous platelet rich plasma. Biomaterials. 2010;**31**:3201-3211. DOI: 10.1016/j.biomaterials.2010.01.038

[35] Eğri S, Eczacıoğlu N. Sequential VEGF and BMP-2 releasing PLA-PEG-PLA scaffolds for bone tissue engineering: I. Design and in vitro tests. Artificial Cells, Nanomedicine, and Biotechnology. 2017;**45**(2):321-329. DOI: 10.3109/21691401.2016.1147454

[36] Wohlfahrt JC, Monjo M, Rnold HJ, Aass AM, Ellingsen JE, Lyngstadaas SP. Porous titanium granules promote bone healing and growth in rabbit tibia peri-implant osseous defects. Clinical Oral Implants Research. 2010;**21**:165. DOI: 10.1111/j.1600-0501.2009.01813.x

[37] Zuchuat J, Berli M, Maldonado Y, Decco O. Influence of chromium-cobalt-molybdenum alloy [ASTM F75] on bone ingrowth in an experimental animal model. Journal of Functional Biomaterials. 2018;**9**(1):2. DOI: 10.3390/jfb9010002

[38] Qian J, Xu W, Yong X, Jin X, Zhang W. Fabrication and in vitro biocompatibility of biomorphic PLGA/nHA composite scaffolds for bone tissue engineering. Materials Science & Engineering. C, Materials for Biological Applications. 2014;**1**(36):95-101. DOI: 10.1016/j.msec.2013.11.047

[39] Sekiya N, Ichioka S, Terada D, Tsuchiya S, Kobayashi H. Efficacy of a poly glycolic acid [PGA]/collagen composite nanofibre scaffold on cell migration and neovascularization in vivo skin defect model. Journal of Plastic Surgery and Hand Surgery. 2013;(6):47, 498-502. DOI: 10.3109/2000656X.2013.788507

[40] Wu C, Zhou Y, Xu M, Han P, Chen L, Chang J, et al. Copper-containing mesoporous bioactive glass scaffolds with multifunctional properties of angiogenesis capacity, osteostimulation and antibacterial activity. Biomaterials. 2013;**34**(2):422-433. DOI: 10.1016/j.biomaterials.2012.09.066

[41] Oughlis S, Lessim S, Changotade S, Bollotte F, Poirier F, Helary G, et al. Development of proteomic tools to study protein adsorption on a biomaterial titanium grafted with poly(sodium styrene sulfonate). Journal of Chromatography B. 2011;**879**:3681-3687. DOI: 10.1016/j.jchromb.2011.10.006

[42] Wichterle O, Lim D. Hydrophilic gels for biological use. Nature. 1960;**185**:117-129. DOI: 10.1038/185117a0

[43] Pasqui D, Torricelli P, De Cagna M, Milena Fini M, Barbucci R. Carboxymethyl cellulose—Hydroxyapatite hybrid hydrogel as a composite material for bone tissue engineering applications. Journal of Biomedical Materials Research. 2014;**102a**(5):1568-1579. DOI: 10.1002/jbm.a.34810

[44] Kinard LA, Dahlin RL, Lam J, Lu S, Lee EJ, Kasper FK, et al. Synthetic

biodegradable hydrogel delivery of demineralized bone matrix for bone augmentation in a rat model. Acta Biomaterialia. 2014;**10**(11):4574-4582. DOI: 10.1016/j.actbio.2014.07.011

[45] Lu L, Mikos AG. The importance of new processing techniques in tissue engineering. MRS Bulletin. 1996;**21**(11):28-32

[46] Mikos AG, Thorsen AJ, Czerwonka LA, Bao Y, Langer R, Winslow DN, et al. Preparation and characterization of poly(l-lactic acid) foams. Polymer. 1994;**35**:1068-1077. DOI: 10.1016/0032-3861(94)90953-9

[47] Thomson RC, Yaszemski MJ, Powers JM, Mikos AG. Fabrication of biodegradable polymer scaffolds to engineer trabecular bone. Journal of Biomaterials Science. Polymer Edition. 1995;**7**(1):23-38

[48] Gomes ME, Ribeiro AS, Malafaya PB, Reis RL, Cunha AM. A new approach based on injection moulding to produce biodegradable starch-based polymeric scaffolds: Morphology, mechanical and degradation behaviour. Biomaterials. 2001;**22**(9):883-889

[49] Mooney DJ, Baldwin DF, Suh NP, Vacanti JP, Langer R. Novel approach to fabricate porous sponges of poly(D,L-lactic-co-glycolic acid) without the use of organic solvents. Biomaterials. 1996;**17**(14):1417-1422

[50] Holy CE, Shoichet MS, Davies JE. Engineering three-dimensional bone tissue in vitro using biodegradable scaffolds: Investigating initial cell-seeding density and culture period. Journal of Biomedical Materials Research. 2000;**51**(3):376-382

[51] Yang S, Leong KF, Du Z, Chua CK. The design of scaffolds for use in tissue engineering. Part I. Traditional factors. Tissue Engineering. 2001;**7**(6):679-689. DOI: 10.1089/107632701753337645

[52] Kim SS, Utsunomiya H, Koski JA, Wu BM, Cima MJ, Sohn J, et al. Survival and function of hepatocytes on a novel three-dimensional synthetic biodegradable polymer scaffold with an intrinsic network of channels. Annals of Surgery. 1998;**228**(1):8-13

[53] Zeltinger J, Sherwood JK, Graham DA, Müeller R, Griffith LG. Effect of pore size and void fraction on cellular adhesion, proliferation, and matrix deposition. Tissue Engineering. 2001;7(5):557-572. DOI: 10.1089/107632701753213183

[54] Hutmacher DW, Schantz T, Zein I, Ng KW, Teoh SH, Tan KC. Mechanical properties and cell cultural response of polycaprolactone scaffolds designed and fabricated via fused deposition modeling. Journal of Biomedical Materials Research. 2001;**55**(2):203-216

[55] Banwart JC, Asher MA, Hassanein RS. Iliac crest bone graft harvest donor site morbidity. A statistical evaluation. Spine. 1995;**20**:1055-1060

[56] Arrington ED, Smith WJ, Chambers HG. Complications of iliac crest bone graft harvesting. Clinical Orthopaedics. 1996;**329**:300-309

[57] Ahlmann E, Patzakis M, Roidis N, Shepherd L, Holtom P. Comparison of anterior and posterior iliac crest bone grafts in terms of harvest-site morbidity and functional outcomes. The Journal of Bone and Joint Surgery. American Volume. 2002;**84A**:716-720

[58] Sikavitsas VI, Temenoff JS, Mikos AG. Biomaterials and bone mechanotransduction. Biomaterials. 2001;**22**(19):2581-2593

[59] Hench LL, Polak JM. Third-generation biomedical materials.

Science. 2002;**295**(5557):1014-1017. DOI: 10.1126/science.1067404

[60] Navarro M, Michiardi A, Castaño O, Planell JA. Biomaterials in orthopaedics. Journal of the Royal Society Interface. 2008;**5**(27):1137-1158. DOI: 10.1098/rsif.2008.0151

[61] Urist MR. Bone: Formation by autoinduction. Science. 1965;**5**(27):1137-1158